DATE DUE	
7033840	4/9/95
857769	MAY 1 4 2000
GAYLORD	PRINTED IN U.S.A.

ENVIRONMENTAL CHANGE AND HUMAN HEALTH

The Ciba Foundation is an international scientific and educational charity (Registered Charity No. 313574). It was established in 1947 by the Swiss chemical and pharmaceutical company of CIBA Limited—now Ciba-Geigy Limited. The Foundation operates independently in London under English trust law. The Ciba Foundation exists to promote international cooperation in biological, medical and chemical research. It organizes about eight international multidisciplinary symposia each year on topics that seem ready for discussion by a small group of research workers. The papers and discussions are published in the Ciba Foundation symposium series. The Foundation also holds many shorter meetings (not published), organized by the Foundation itself or by outside scientific organizations. The staff always welcome suggestions for future meetings. The Foundation's house at 41 Portland Place, London W1N 4BN, provides facilities for meetings of all kinds. Its Media Resource Service supplies information to journalists on all scientific and technological topics. The library, open five days a week to any graduate in science or medicine, also provides information on scientific meetings throughout the world and answers general enquiries on biomedical and chemical subjects. Scientists from any part of the world may stay in the house during working visits to London.

The European Environmental Research Organisation (EERO) was established in 1987 by a group of scientists from seven European countries. The members, now drawn from 14 countries, are scientists elected in recognition of their outstanding contributions to environmental research or training. EERO became fully operational in 1990 as a non-profit making, non-political foundation, with its offices at the Agricultural University of Wageningen, Generaal Foulkesweg 70, 6703 BW Wageningen, The Netherlands. EERO seeks to strengthen environmental research and training in Europe and neighbouring countries, from the Atlantic to the Urals, and from Scandinavia to the Mediterranean. It does so by promoting the most effective use of intellectual and technological resources and by supporting new and emerging scientific opportunities. EERO also provides a resource for authoritative but independent scientific assessment of environmental issues and institutions. Particular objectives are: to develop the potential of promising younger scientists by providing opportunities for collaborative research and training; to bring together participants from the relevant basic sciences to form interdisciplinary environmental research groups; to catalyse the formation and evolution of international networks to exploit emerging opportunities in environmental research; to facilitate the rapid spread in Europe of new theoretical knowledge from all parts of the world; to provide training in new developments in environmental science for those in government, industry and the research community; and to make independent scientific assessments of environmental problems and institutions, as a basis for advice to government or industry.

EUROPEAN ENVIRONMENTAL
RESEARCH ORGANISATION

ENVIRONMENTAL CHANGE AND HUMAN HEALTH

A Ciba Foundation Symposium
jointly with the European Environmental Research Organisation

A Wiley-Interscience Publication

1993

JOHN WILEY & SONS

Chichester · New York · Brisbane · Toronto · Singapore

©Ciba Foundation 1993

Published in 1993 by John Wiley & Sons Ltd
Baffins Lane, Chichester
West Sussex PO19 1UD, England

Other Wiley Editorial Offices

John Wiley & Sons, Inc., 605 Third Avenue,
New York, NY 10158-0012, USA

Jacaranda Wiley Ltd, G.P.O. Box 859, Brisbane,
Queensland 4001, Australia

John Wiley & Sons (Canada) Ltd, 22 Worcester Road,
Rexdale, Ontario M9W 1L1, Canada

John Wiley & Sons (SEA) Pte Ltd, 37 Jalan Pemimpin #05-04,
Block B, Union Industrial Building, Singapore 2057

Suggested series entry for library catalogues:
Ciba Foundation Symposia

Ciba Foundation Symposium 175
viii + 274 pages, 47 figures, 16 tables

Library of Congress Cataloging-in-Publication Data
Environmental change and human health / editors, John V. Lake,
    Gregory R. Bock, and Kate Ackrill.
        p.   cm.—(Ciba Foundation symposium; 175)
        "Symposium on Environmental Change and Human Health, held in
    collaboration with the European Environmental Research Organisation,
    at the No 1 in't Bosch Hotel, Wageningen, The Netherlands on 1–3
    September 1992."
        "A Wiley–Interscience publication."
        Includes bibliographical references and index.
        ISBN 0 471 93842 4
        1. Environmental health—Congresses.   I. Lake, J. V.   II. Bock,
    Gregory.   III. Ackrill, Kate.   IV. European Environmental Research
    Organisation.   V. Symposium on Environmental Change and Human Health
    (1992 : Wageningen, Netherlands).   VI. Series.
    RA565.A2E556   1993
    616.9'8—dc20   93-15262
                                                                              CIP

British Library Cataloguing in Publication Data
A catalogue record for this book is
available from the British Library

ISBN 0 471 93842 4

Phototypeset by Dobbie Typesetting Limited, Tavistock, Devon.
Printed and bound in Great Britain by Biddles Ltd, Guildford.

Contents

Participants

Y. Avnimelech   Office of the Chief Scientist, State of Israel, Ministry of the Environment, PO Box 6234, Jerusalem 91061, Israel

M. Bobrow   Division of Medical and Molecular Genetics UMDS, Paediatric Research Unit, 8th Floor, Guy's Tower, Guy's Hospital, London Bridge, London SE1 9RT, UK

D. J. Bradley   Department of Epidemiology & Population Sciences and Ross Institute, London School of Hygiene and Tropical Medicine, Keppel Street, Gower Street, London WC1E 7HT, UK

F. A. M. de Haan   Department of Soil Science and Plant Nutrition, Wageningen Agricultural University, PO Box 8005, Dreijenplein 10, 6700 EC Wageningen, The Netherlands

J. H. Edwards   Genetics Laboratory, Department of Biochemistry, University of Oxford, South Parks Road, Oxford OX1 3QU, UK

P. Elliott   Environmental Epidemiology Unit, Department of Public Health & Policy, London School of Hygiene and Tropical Medicine, Keppel Street, Gower Street, London WC1E 7HT, UK

J. G. A. J. Hautvast   Department of Human Nutrition, Wageningen Agricultural University, Bomenweg 2, PO Box 8129, NL-6700 EV Wageningen, The Netherlands

M. R. Hoffman   Department of Environmental Engineering Science, W. M. Keck Laboratories 138-78, California Institute of Technology, Pasadena, CA 91125, USA

W. P. T. James   Rowett Research Institute, Greenburn Road, Bucksburn, Aberdeen AB2 9SB, UK

J. Klein   GSF-Forschungszentrum für Umwelt und Gesundheit, Ingolstädter Landstrasse 1, D-8042 Neuherberg, Germany

J. Kleinjans Department of Health Risk Analysis & Toxicology, University of Limburg, PO Box 616, 6200 MD Maastricht, The Netherlands

J. V. Lake EERO, Generaal Foulkesweg 70, PO Box 191, 6700 AD, Wageningen, The Netherlands

D. W. Lincoln MRC Reproductive Biology Unit, University of Edinburgh Centre for Reproductive Biology, 37 Chalmers Street, Edinburgh EH3 9EW, UK

T. A. Mansfield Institute of Environmental & Biological Studies, University of Lancaster, Bailrigg, Lancaster LA1 4YQ, UK

H. Oeschger Institute of Physics, University of Berne, Sidlerstrasse 5, CH-3012 Berne, Switzerland

R. Rabbinge Department of Theoretical Production Ecology, Wageningen Agricultural University, PO Box 430, 6700 AK Wageningen, The Netherlands

R. W. Sutherst CSIRO Division of Entomology, Cooperative Research Centre for Tropical Pest Management, Gehrmann Laboratories, University of Queensland, Brisbane, Queensland 4072, Australia

U. Vogel Division of Environmental Process Technology, Ciba-Geigy Werke Schweizerhalle AG, Verfahrenstechnik Umweltschulz, Bau 2090.3.06, CH-4133 Schweizerhalle, Switzerland

A. J. B. Zehnder Swiss Federal Institute of Technology (ETH) Zürich, Federal Institute for Water Resources and Water Pollution Control (EAWAG), Ueberlandstrasse 133, CH-8600 Dübendorf, Switzerland

Introduction

Alexander J. B. Zehnder

Swiss Federal Institute of Technology (ETH) Zürich, Federal Institute for Water Resources and Water Pollution Control (EAWAG), Ueberlandstrasse 133, CH-8600 Dübendorf, Switzerland

A meeting on Environmental Change and Human Health is timely. Almost every day news reaches us of famines, of areas newly heavily polluted where the people's health is affected by contaminated air or drinking water, of the degradation of stratospheric ozone, and of the increase in CO_2 in the atmosphere and consequent higher temperatures on earth. Scenarios about environmental and human future are painted, ranging from frightening to reassuring. Besides these scenarios are predictions of an explosive growth in world population, which raise questions about how we are to feed and provide shelter for this growing population without exhausting fundamental resources.

Against this background, the Ciba Foundation and the European Environmental Research Organisation (EERO) agreed to hold a joint symposium on various aspects of the changing environment and human health. The combination of these two organizations was perfect—the Ciba Foundation with its special emphasis on advancing cooperation in medical, chemical and biological research, and EERO with its programme to promote the most effective use of intellectual and technological resources and to stimulate interdisciplinary collaboration in environmental research and on environmental issues.

This symposium brings together scientists from a range of disciplines, including physics, chemistry, biology, agriculture and medical science, to discuss the various questions related to the theme of environmental change and human health. The discussions should be used not only to define current knowledge better, but also—and I consider this to be more important—to find new solutions, and where new solutions cannot be identified, to highlight crucial research areas and directions. Ladies and gentlemen, let us start with our work.

CO$_2$ and the greenhouse effect: present assessment and perspectives

Hans Oeschger

Institute of Physics, University of Berne, Sidlerstrasse 5, CH-3012 Berne, Switzerland

Abstract. Our present knowledge on the increasing greenhouse effect is based on the 1990 assessment of the Intergovernmental Panel on Climate Change (IPCC) and its 1992 supplement. Model predictions suggest that a doubling of atmospheric CO$_2$ concentration would increase global temperature by 2–4 °C. Because time is needed for the upper layers of the oceans to warm up, there is a delay between the realized increase and the estimated equilibrium increase. The 0.5 °C increase in global temperature over the past 100 years is in accordance with model predictions but also within the range of natural variability. The IPCC's assessment suggests that if fossil fuel consumption continues according to a 'business as usual' scenario, global temperature will increase at 0.3 °C per decade; droughts, flooding and storms would become more frequent and more severe, and sea-level would continue to rise. Analysis of gas in ice cores from Antarctica and Greenland provides estimates of CO$_2$ concentrations in pre-industrial ages; accurate measurement of atmospheric CO$_2$ began in 1958. Atmospheric CO$_2$ concentration has increased from 280 p.p.m. in AD 1800 to 355 p.p.m. at present. Between 1945 and 1973 global emissions of CO$_2$ increased at a rate of 4.4% per year; after 1973 the rate of increase decreased to 1.6% per year. This change permits a test of the CO$_2$ model. Reconstructed CO$_2$ emission agrees within less than 10% with estimated fossil fuel-generated CO$_2$ emission; contributions to CO$_2$ emission from non-fossil fuel sources must be smaller than assumed previously. There are strong indications that in the past changes in atmospheric greenhouse gas concentrations inherent to the glacial–interglacial cycles did play an important climatic role. For example, they were probably responsible for interhemispheric coupling during the major climatic changes. Measures which might stabilize the greenhouse effect include energy conservation and improved energy efficiency, a transition to hydrogen rather than carbon as a source of fuel, and reforestation.

1993 Environmental change and human health. Wiley, Chichester (Ciba Foundation Symposium 175) p 2–22

Roughly 30% of the solar energy falling onto the earth is reflected directly back into space; 20% is absorbed by the atmosphere and 50% by the ground (Fig. 1). With an average temperature of 15 °C (288 K) the surface of the earth and the air near the ground emit infrared radiation. Whereas the cloud-free atmosphere is almost transparent for the incoming solar radiation, most of the emitted

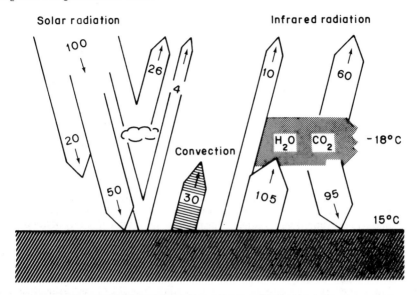

FIG. 1. The energy budget of the earth. Solar irradiation = 100 units. *Left*, solar radiation; *right*, infrared radiation. The infrared radiation emitted from the atmosphere to the surface (95 units) exceeds the absorbed solar radiation (70 units).

infrared radiation is absorbed by water vapour, CO$_2$ and trace gases. Part of the absorbed energy is re-emitted back to the ground. This increases the temperature of the earth's surface and the air near the ground above that produced by the absorption of solar energy alone.

This warming of the ground is called the greenhouse effect. For a radiation equilibrium to be established at the top of the atmosphere, the temperature of the surface and the near-ground atmosphere must be high enough to result in emission of infrared radiation corresponding to 115% of the solar energy falling on the earth, in spite of the fact that only 70% of the solar energy is absorbed. The temperature of the near-ground air is 15 °C, 33 °C higher than the planetary temperature (as derived from the emission at the top of the atmosphere).

Satellite observations have allowed us to determine the greenhouse effect on Venus and Mars; the temperature increases found are compatible with the greenhouse theory.

Changes of the greenhouse effect and their consequences

If the atmospheric concentrations of CO$_2$ and other greenhouse gases increased, initially less infrared radiation would be released to space and the radiation balance at the top of the atmosphere would be disturbed. The surfaces of the continents and the oceans, as well as the low air masses, would then become warmer and emit more and more infrared radiation until a new radiation equilibrium became established.

Detailed calculations show that for a doubling of CO_2 (or the radiative equivalent) the emission of infrared radiation to space is reduced by about $4\,W/m^2$. To compensate for this loss in emission the surface of the earth and the near-ground air layers need to heat up by $1.2\,°C$. This temperature increase is termed the primary forcing for a CO_2 doubling. It is a well-assessed quantity.

Feedbacks

A change in the temperature of the earth's surface induces a series of feedback effects which amplify the primary forcing. These feedbacks are summarized in (1). The absolute humidity increases and reinforces the

$$\Delta T = 1.2\,°C \times \frac{1}{1 - 0.4(H_2O - Vapour) - 0.1(\text{Snow, Ice}) - (0.2 \text{ to } -0.1\,[\text{Clouds}])}$$

$$(1)$$

greenhouse effect, and the back-scattering of solar radiation is amplified because of the decrease in the snow and ice cover. Both these feedback effects are positive. Less is known about the feedback effect of changing cloud cover; it can be positive (clouds in higher, colder layers which emit less infrared radiation) or negative (higher water content leading to brighter clouds).

Equation (1) suggests there would be a temperature increase (ΔT) of $2\text{–}4\,°C$ in response to a doubling of CO_2. The estimates obtained with general circulation models (GMCs) are also in this range (Intergovernmental Panel on Climate Change [IPCC]; Houghton et al 1990, 1992).

Equilibrium and realized temperature increases

The lower atmospheric layers are strongly coupled to the surface of the oceans and, because of ocean mixing for the present rate of increase of the greenhouse forcing, the equivalent of an ocean layer several hundred metres deep needs to be warmed up. (For comparison, the heat capacity of the atmosphere corresponds to that of an ocean layer approximately two metres deep). For this reason, the realized atmospheric temperature increase lags behind the equilibrium temperature increase, and for the present rate of increase corresponds to only 60% of the equilibrium value. In addition, the warming is expected to show an ocean–continent asymmetry, the temperature increase over the oceans lagging behind that over the continents.

Sea-level rise

The temperature increase of a relatively thick ocean layer leads, because of thermal expansion, to a rise in sea-level. The thermal expansion coefficient depends on the temperature of the water. It is significantly greater for warm surface water of low latitudes than for the cold water which sinks at high northern latitudes of the Atlantic ocean to the deep sea and then flows to the south. In estimating the sea-level rise caused by thermal expansion it is therefore important to consider how the excess heat penetrates the ocean. A substantial part of the excess heat is thought to be transported with cold water into deeper ocean layers.

Even more difficult than estimating the sea level rise caused by thermal expansion is predicting the effects of melting of continental ice. It is believed that the ablation of ice in the Greenland ice sheet will increase. Accumulation in the interior part is also expected to increase because of increased precipitation, which should compensate for part of the loss of land ice in the marginal zones. For the Antarctic ice sheet, an initial increase in accumulation seems likely. In the longer term, over centuries, instability in parts of the ice sheet could have drastic consequences for sea-levels.

Biotic feedback effects

It is difficult to estimate the feedback effects on climate resulting from changes in the sources and sinks of greenhouse gases. Important information can be obtained by using ice core analysis to reconstruct the history of atmospheric greenhouse gas concentrations. The atmospheric concentrations of CH_4 and N_2O are positively correlated with global temperature, a strong indication of a positive feedback effect in the context of global warming. The relationship between CO_2 and temperature is more complex. Hitherto, biotic feedback has not been quantitatively considered in the estimates of future climate.

Increases in the greenhouse effect

The temperature increase of the past hundred years

The trend in global temperature over the last hundred years is shown in Fig. 2. The average yearly temperatures show variations of about ± 0.2 °C around the five-year average temperatures, but the latter also do not show a smooth trend. The variations can partly be explained by volcanic eruptions (Mount Agung in 1962 and El Chichón in 1982), which lead to increased atmospheric turbidity. Also, the El Niño southern oscillation phenomenon leads to small global temperature deviations, and also possibly smaller variations in solar luminosity. The eight warmest years of the entire observation period were in the 1980s and the beginning of the 1990s.

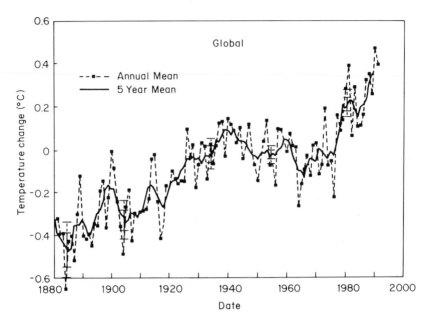

FIG. 2.   Global temperature change; annual (-----) and five-year (——) means. Figure kindly provided by J. Hansen and H. Wilson, National Aeronautics and Space Administration Goddard Institute for Space Studies.

The observed increase in temperature of 0.5 °C over the past 100 years is in accordance with what one would expect on the basis of model calculations, but it is also still within the range of natural variability. The fact that one cannot yet unambiguously attribute this warming to the increasing greenhouse effect does not mean that scientists do not agree that the identified increase of the greenhouse forcing will lead to global warming in the range given above.

The increase in atmospheric CO_2 since AD 1800

Continuous high precision measurements of atmospheric CO_2 began in only 1958, at Mauna Loa, Hawaii, and at the South Pole. As shown in Fig. 3, CO_2 has increased from 315 p.p.m. to 355 p.p.m. at both stations. The pronounced seasonal variations, as observed at Mauna Loa, appear to be strongly dampened at the South Pole. Figure 4 shows in schematic form the exchange fluxes of atmospheric CO_2 with the carbon in the ocean and in vegetation and soils. During one year about a quarter of the CO_2 in the atmosphere is exchanged with the other reservoirs. This raises two questions: why can this system not buffer the relatively small CO_2 input generated by human activities, and do these large exchange fluxes between the reservoirs always balance each other out or are there periods of disequilibrium? If there are periods of

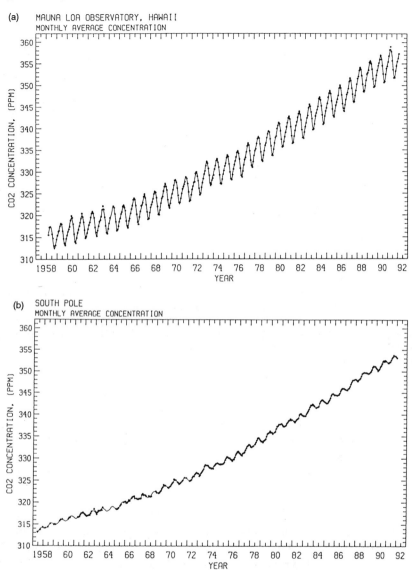

FIG. 3.   Atmospheric CO$_2$ concentrations (p.p.m. dry air) from 1958 to 1992 measured at (a) Mauna Loa Observatory, Hawaii, and (b) at the South Pole. The dots indicate monthly average concentration. Figures kindly provided by C. D. Keeling and T. P. Whorf (Keeling et al 1989, 1987). Mauna Loa Observatory is operated by US National Oceanic and Atmospheric Administration (NOAA). The measurements were obtained in a cooperative programme between NOAA and the Scripps Institution of Oceanography.

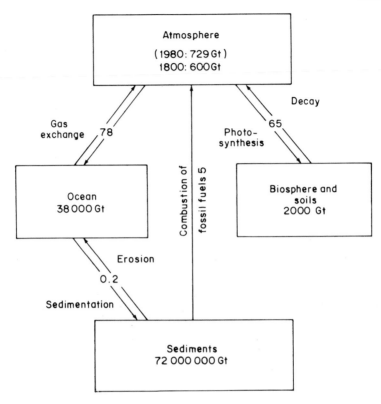

FIG. 4.   The exchange of CO_2 between the atmosphere, the biosphere, oceans and sediments. CO_2 contents of the reservoirs are indicated in gigatonnes (Gt) C per year.

disequilibrium, could the recent CO_2 increase not reflect such a disequilibrium? To give convincing answers to these questions, one needs to reconstruct the history of atmospheric CO_2 concentrations from the pre-industrial age. It is possible to do this because air bubbles in natural ice constitute physically occluded samples of the ancient atmosphere. Atmospheric CO_2 concentrations of the past 200 years have been reconstructed from analysis of samples from an ice core retrieved at Siple Station, Antarctica (Neftel et al 1982, Friedli et al 1986). The concentrations reconstructed from the uppermost ice layers overlap with the Mauna Loa record (Keeling et al 1989). In the second part of the 18th century the CO_2 concentration was about 280 p.p.m. Analysis of another core at South Pole Station showed that during the last, pre-industrial, millenium the concentration of CO_2 was in the range 280–290 p.p.m. These data allow us to conclude unequivocally that the increase evident in the 19th century (compare also the direct atmospheric CO_2 measurements; Fig. 3) is a consequence of human impact on the biomass and of the later rapidly increasing use of fossil energy.

The prognoses of the assessment of the IPCC

Some of the important conclusions of the IPCC assessment are given below. The IPCC has assessed the relative contributions of anthropogenic greenhouse gases to the greenhouse forcing in 1990. The most important of these is CO_2, contributing 55%. The other gases and their relative contributions are: 11 and 12 chlorofluorocarbons (CFCs), 17%; other CFCs, 7%; nitrous oxide, 6%; and methane, 15% (Houghton et al 1990). The contribution of ozone is difficult to assess and is not considered. Figure 5a shows the assumed future increases of CO_2, CH_4 and CFC-11, on which the IPCC calculations of future climate forcing are based. The 'business as usual' scenario is coal-intensive, and under this scenario CFC emissions would be partly reduced. Under scenario B there would be a transition to fuels with higher H:C ratios (predominantly natural gas). Scenarios C and D involve a shift to renewable energies and nuclear energy. The increases of global temperature and sea-level predicted under the different scenarios are shown in Fig. 5b. If emissions continue at the present rate there would be an increase in global temperature of 0.3 °C per decade, whereas under scenario D the temperature increase would stabilize at 2 °C by the year 2100.

Temperature and greenhouse gas concentrations of the past 150 000 years

Figure 6 shows changes in atmospheric CO_2 and CH_4 concentrations and (Antarctic) temperature over the past 150 000 years. The deviations shown in temperature are roughly double those in global temperature derived from independent information. As indicated above, global climatic changes are coupled with changes in greenhouse gas concentrations. There is an amazingly good correlation between temperature and CH_4 concentrations, which varied by a factor of up to two over this period. Increasing global temperature probably increases the area of wetlands and thus leads to increased anaerobic decomposition and methane sources; during warmer periods methane may have been released in large quantities from thawing permafrost. The CO_2 concentrations also vary in parallel with temperature; the transitions from the coldest glacial temperatures to the warm interglacial ones are accompanied by drastic increases in CO_2. However, the relationship is more complicated; a kind of hysteretic effect is evident. This is probably caused by emission of CO_2 during (and after) climatic deterioration resulting from a decrease of terrestrial biomass, and by the uptake of CO_2 by a growing biomass during (and after) a considerable climatic improvement. Changes in ocean chemistry are probably the main driving force for the changes in CO_2. Particularly important are processes in the ocean surface in which the partial pressure of CO_2 is strongly affected by the biological pump (the sinking of organic particles to greater depth). Changes in the biological pump could result in large changes in

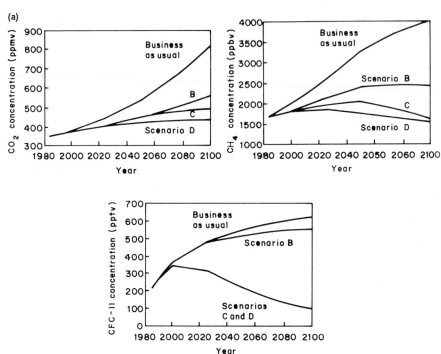

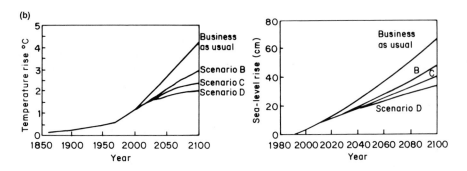

FIG. 5. The prognoses of the Intergovernmental Panel on Climate Change (IPCC).
(a) The predicted atmospheric concentrations of *top left*, CO_2 (in parts per million
by volume), *top right*, CH_4 (parts per billion by volume) and *middle*, chlorofluoro-
carbon (CFC)-11 (parts per trillion by volume) to the year 2100 under four different
scenarios: the 'business as usual' scenario; scenario B, with a transition to fuels of higher
H:C ratio; scenario C, with a transition to renewable energy forms and nuclear
energy in the second half of the next century; and scenario D, involving renewable
energy and use of nuclear energy in the first half of the next century. (b) The increase
in temperature (*bottom left*) and sea level (*bottom right*) predicted under the four
scenarios.

atmospheric CO_2: were all the nutrients in the surface ocean to be used up by organic processes, the partial pressure of CO_2 (in the water and in the atmosphere) would decrease to only 150 p.p.m. If there were no biological activity in the ocean surface, the partial pressure of CO_2 would rise to about 500 p.p.m.

A question often posed is, which was cause and which effect—the shift in temperature or that in the greenhouse gases (mainly CO_2) over the past 160 000 years? Present knowledge suggests the changes of temperature and greenhouse gases are due to interactive processes inherent to the climate system which lead to a positive feedback effect.

According to the theory of ice ages, one can consider the glaciations of the northern hemisphere as a result of changes in the orbital elements which induced ice cover over large areas of the continents, leading to a significantly enhanced albedo. It is, however, very difficult to understand why the temperature oscillations in the southern hemisphere were in parallel with those in the northern hemisphere. Global greenhouse gas variations might provide an explanation for this. They generate a forcing of 3 W/m^2, about three-quarters of that provided by a doubling of CO_2, during the transition from the coldest part of a glaciation to the following interglacial period. It thus seems probable that the climate of the past million years was strongly influenced by changes of the greenhouse effect (of the order of magnitude of those caused by humans). It is obvious that the interhemispheric coupling during a glaciation cycle is worthy of study; the changing greenhouse effect might have played a major role.

Further lessons from the past

How will climate change as a consequence of human activity, and how stable will future climate be? The analysis of ice cores from Greenland shows that between 80 000 BP and 30 000 BP the climate of the north Atlantic region oscillated more than a dozen times back and forth between a cold and a mild state, probably because of changes in ocean circulation patterns. In our present interglacial era climatic variability has been much less pronounced. How variable was climate during the last interglacial period, 120 000 to 130 000 years ago, when temperature was about 2 °C higher and sea-level about 5 m higher than at present? Does global warming lead to a more stable climate or to a period with strong climatic fluctuations? An indication of the answers to these questions will be obtained soon from the analysis of the newly drilled ice core from the summit of the Greenland ice sheet. One could imagine that in a warmer climate there could be more states of the earth system, and more transitions between them. This would be in disagreement with the observed strong climatic variability during the last interglacial era, but during a glaciation instabilities of the land ice could lead to an enhanced variability of the earth system.

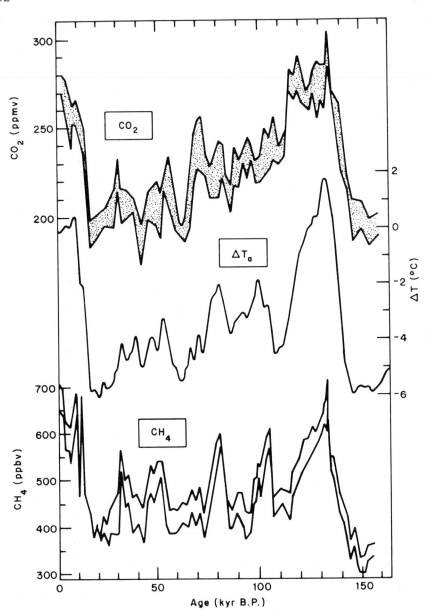

FIG. 6. Ice records from Vostok, Antarctica, for methane (Chappellaz et al 1990), deviations in isotope temperature above the inversion layer (as a difference from the present value of about $-40\,°C$) and CO_2 (Barnola et al 1987). The CH_4 and CO_2 curves have been adjusted to fit the timescale given (taking into consideration gas occlusion time). Taken from Oeschger (1991), with permission. ©Cambridge University Press.

CO$_2$ emissions and their increase over the last few decades

Figure 7 shows a logarithmic plot of the estimated global emissions of CO$_2$ over the last four decades. From 1950 to 1973 the growth rate was essentially constant at 4.4% per year; it then decreased to 1.6% per year, on average. The present emission of CO$_2$ resulting from consumption of fossil fuels is about 6 Gt C per year. Had the high growth rate continued, this emission would have risen to 10 Gt C per year. This decrease in the rate of increase of CO$_2$ emissions is very important for attempts to stabilize atmospheric CO$_2$ concentration. If one aimed to stabilize atmospheric CO$_2$ at 420 p.p.m., i.e. 150% of the pre-industrial concentration, one would, according to the present models for CO$_2$ uptake by the ocean, need to reduce emissions from 6 Gt C per year to 3 Gt C per year by the year 2050, that is, by about 50%. Had emissions continued to increase at 4.4% per year, it would be necessary to reduce emissions from 10 Gt C per year to 2 Gt C per year, that is, by 80%. This clearly demonstrates that early action on control of CO$_2$ emissions is necesary, in

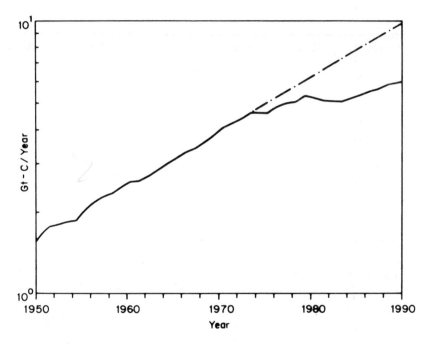

FIG. 7.   Logarithmic plot of estimated CO$_2$ emission resulting from consumption of fossil energy from 1950 to 1990. Up to 1973, the rate of increase in emissions was 4.4% per year; between 1973 and 1990 the average growth rate was 1.6% per year (F. Joos, H. Oeschger & U. Siegenthaler, unpublished work). The dotted line shows the amount of CO$_2$ that would have been emitted had fossil fuel consumption continued to increase at a rate of 4.4% per year.

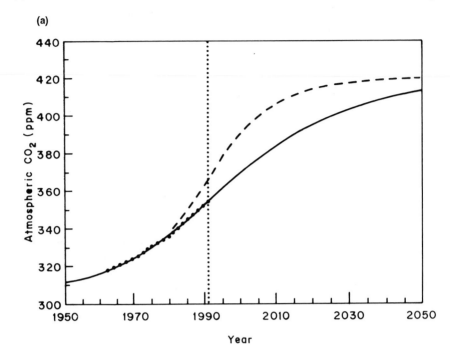

(a)

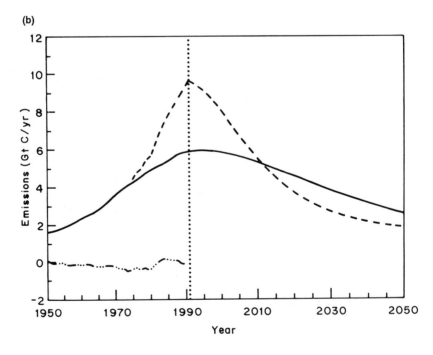

(b)

contradiction to the opinion of some that action is not urgent because CO$_2$ is a long-term problem!

The decrease in the rate of increase of CO$_2$ emission permits an interesting test of the CO$_2$ model and our present understanding of the perturbation of the carbon cycle by human activities. Figure 8 shows the measured atmospheric CO$_2$ concentration from 1945 to 1990. With a carbon cycle model describing the uptake of excess CO$_2$ by the ocean (calibrated on the basis of data about ^{3}H and ^{14}C produced by nuclear weapons and about CFCs in the ocean) the total (from fossil fuel consumption and the impact on the biomass) CO$_2$ emission history can be deconvolved from the observed CO$_2$ increase. The pattern of emission so generated agrees within better than 10% with the estimated fossil CO$_2$ emission over the entire period. The difference is small in comparison with the overall uncertainties. This result suggests that (i) the atmospheric CO$_2$ data and the estimates of emissions from fossil fuel consumption are in close agreement with reality; (ii) the CO$_2$ model fairly precisely describes CO$_2$ uptake by the oceans, including the shift to greater uptake during the period of lower growth; and (iii) the non-fossil contribution to CO$_2$ emissions is smaller than generally assumed. There is no indication of a large missing sink.

Measures to stabilize the greenhouse effect

Measures intended to reduce CO$_2$ emissions involve saving energy, increasing efficiency of energy use, a transition from carbon to hydrogen as a source of fuel, development and use of non-fossil energies and reforestation.

To what extent would reforestation compensate for CO$_2$ emission? Assume that during the next 40 years 10 million ha forest are planted every year. In 2030 40 million ha (100 times the area of Switzerland) of new forest would absorb 1 Gt C. The process would continue for about 40 to 100 years, until the trees had reached an equilirbium. According to this scenario, up to 2030 20 Gt C

FIG. 8 *(opposite)*.   (a) Atmospheric CO$_2$ concentrations measured at Mauna Loa, Hawaii (dots); the solid line is fitted to the data up to 1990 and then extrapolated up to the value of 420 p.p.m. The dashed line shows the increase that would occur if CO$_2$ emissions had continued to increase by 4.4% per year, then levels off at 420 p.p.m. (b) The solid line represents the deconvolved measured CO$_2$ increase (see solid line in [a]). The dashed/dotted line is the difference between the deconvolved emission and the estimated emission. The dashed line reflects the total emission that would have resulted had the high growth continued up to 1990. Between 1990 and 2050 the solid and the dashed lines represent the emissions corresponding to the projected atmospheric CO$_2$ concentrations shown in part (a) (F. Joos, H. Oeschger & U. Siegenthaler, unpublished work).

and in 100 years time some 80 Gt C could be fixed. This accumulation of CO_2 would, however, correspond to only 5–10% of the CO_2 emissions from fossil fuel consumption predicted under the IPCC's business as usual scenario.

In the past, various types of energy replaced one another (wood, carbon, petrol, natural gas). In the beginning of the 1970s it was expected that nuclear energy would, to a significant extent, replace fossil energy, but nuclear energy was considered to be too risky in most countries. An exception is France, where at present 80% of electrical energy is produced by nuclear power. It is interesting to compare France's CO_2 emissions with those of other Western countries which developed nuclear energy to a lesser extent, such as the USA. The difference is striking. CO_2 emissions from fossil fuel consumption in the USA have remained essentially constant, whereas from 1977 to 1987 they decreased in France, essentially because of the intensive use of nuclear energy, by 29%. This clearly contradicts some intellectual constructions which claim that nuclear energy would lead to an increase in CO_2 emissions.

Conclusions

If the emissions of greenhouse gases continue according to the IPCC's 'business as usual' scenario, global temperature in the next century would increase by as much as 0.3 °C per decade, a rate of change greater than any over the past 10 000 years. Droughts and flooding would become more frequent and more violent. Regions important for growing food for the world's population would become infertile, and others infertile at present would become fertile. Storm activity would become more intense and unexpected developments could have negative impacts on the survival of people in certain parts of the world. Sea-level would continue to rise.

Over the last years and months, doubts have been raised about the results of global change science and the validity of the IPCC's assessments. Science needs a continuous questioning of its results in order to make progress. However, the arguments and the criticisms raised are largely of mediocre quality and cannot put in doubt the essential conclusions of the scientific community who have been involved in this research for decades. It is sometimes argued that when we estimate the sum of the positive and the negative changes we may come to the conclusion that global warming would be beneficial; even if this were the case, would it be possible to make use of the positive developments in time to compensate for the negative ones? Already it is not possible for those suffering from famine to profit from the excess of food production in other areas. Divergence in living conditions will increase further, intensifying the problems of population migration. There is an urgent need to slow down global climatic change, but this is possible only if we have a sound economy. We have to find a way which minimizes both the stress on the environment and that on the economy. Technology and economy should take the lead in a move towards sustainable development.

Acknowledgements

Thanks are due to my colleagues in the field of the physics and chemistry of climate and environment, including C. D. Keeling, H. E. Suess, W. C. Broecker, C. C. Langway, W. Dansgaard and C. Lorius.

References

Barnola JM, Raynaud D, Korotkevich VS, Lorius C 1987 Vostok ice core: a 160,000 year record of atmospheric CO$_2$. Nature 329:408–414
Chappellaz J, Barnola JM, Raynaud D, Korotkevich YS, Lorius C 1990 Ice-core record of atmospheric methane over the past 160,000 years. Nature 345:127–131
Friedli H, Loetscher H, Oeschger H, Siegenthaler U, Stauffer B 1986 Ice core record of the ^{13}C/^{12}C ratio of atmospheric CO$_2$ in the past two centuries. Nature 324:237–238
Houghton JT, Jenkins GK, Ephraums JJ (eds) 1990 IPCC scientific assessment. Cambridge University Press, Cambridge
Houghton JT, Callander BA, Varney SK (eds) 1992 Supplementary report of the IPCC scientific assessment. Cambridge University Press, Cambridge
Keeling CD, Moss DJ, Whorf TP 1987 Measurements of the concentration of atmospheric carbon dioxide at Mauna Loa Observatory, Hawaii, 1958–1986. Final report for the Carbon Dioxide Information and Analysis Center, Martin-Marietta Energy Systems Inc, Oak Ridge, TN
Keeling RB, Bacastow AF, Carter SC et al 1989 A three dimensional model of atmospheric CO$_2$ transport based on observed winds. I. Analysis of observational data. In: Peterson DH (ed) Aspects of climate variability in the Pacific and the Western Americas. American Geophysical Union, Washington (Geophys Monogr 55) p 165–236
Neftel A, Oeschger H, Schwander J, Stauffer B, Zumbrunn R 1982 Ice core measurements give atmospheric CO$_2$ content during the past 40,000 yr. Nature 295:220–223
Oeschger H 1991 Paleodata, paleoclimates and the greenhouse effect. In: Jäger J, Ferguson HL (eds) Climate change: science, impacts and policy. Cambridge University Press, Cambridge (Proc Sec World Clim Conf) p 215–224

DISCUSSION

Edwards: Do ruminants make any appreciable contribution to methane emissions?

Oeschger: Detailed figures do not exist, but I think they are estimated to contribute about a quarter or so of the total production.

Hoffmann: The primary source of methane is rice paddy-fields. Anaerobic decomposition there provides the largest fraction of methane flux into the atmosphere. About 10 years ago the ruminant contribution was thought to be quite high, but after fairly careful and serious analyses in different locations around the globe, it was realized that ruminants contribute much less than previously thought (Cirerone & Oremland 1988).

de Haan: Waste disposal sites produce a substantial amount of methane. Is this negligible in comparison with other sources?

Hoffmann: Certainly. If you consider the acreage devoted to rice growing you realize this is the dominant global source.

Avnimelech: To what extent do wetlands contribute?

Hoffmann: They contribute about 115 teragrammes (Tg) per year to the methane flux. The people who do research on global change have gone out into the field and measured methane fluxes from each of these environments, and after about 10 years of this type of research they have been able to get pretty good measurements of the fluxes from different environmental settings and different situations.

Oeschger: The contribution from natural gas leakage, on the basis of carbon isotope studies, appears to be of the order of 20%.

Edwards: If the carbon atom in methane is oxidized to generate CO_2, is that better or worse for the environment? You pointed out that methane and CO_2 are both undesirable in excess. If you turn one into the other, which you do if you burn methane, is that good or bad?

Oeschger: Methane has an estimated mean life of about 12 years in the atmosphere, where it is burnt to CO_2 through chemical processes anyway. Per molecule, CH_4 has a much higher greenhouse effect than CO_2. Thus, by burning CH_4 into CO_2 one avoids this period (12 years) during which it has a high greenhouse effect.

Hoffmann: The concentration of methane in the atmosphere, 1.7 p.p.m. by volume, is much lower than that of CO_2, 350 p.p.m.v., so its relative contribution to CO_2 loading in the atmosphere is quite small.

Mansfield: Professor Oeschger, are you convinced that we know enough about the future production of N_2O? We are adding soluble nitrogen compounds to the atmosphere all the time in the form of NO_x and ammonia, which then get taken up by vegetation and deposited in leaf litter, and will contribute to future microbiological activity. Isn't it likely that by putting NO_x and ammonia into the atmosphere, which do not by themselves contribute much to the greenhouse effect, we might be appreciably accelerating the future production of N_2O? This might be some distance into the future, with a lag of 10 or 20 years or more, but that could be very important in view of the role of N_2O within the atmospheric window.

Oeschger: I agree. It's difficult to estimate what the increase will be. The problem of N_2O sources is very complex.

Lake: We have been reducing nitrogen from the atmosphere to produce nitrogen fertilizers for world agriculture. We will discuss population growth later, but it's worth pointing out here that to feed the extra people we will be combining atmospheric nitrogen in a form which could contribute to the greenhouse effect when later it's released .

Mansfield: That's essentially what's going on. On top of that is the production of NO_x during combustion processes.

Zehnder: How much N_2O or NO_x comes from agriculture and how much from combustion processes?

Mansfield: Over much of Western Europe, the rate of wet and dry deposition of nitrogen is about 40 kg/ha/year. About 60% of that comes from agricultural activity, mainly in the form of ammonia, and 40% is NO_x deposition, which originates mainly from combustion processes. 40 kg/ha/year is a huge amount. In intensive agriculture you would expect to be adding about 100 kg/ha/year nitrogen as fertilizer. What is often missed by people discussing this issue is that the 40 kg/ha/year is on *every* hectare of land, not only on the agricultural land, but also on every road surface, every roof-top, every garden. In populated areas like The Netherlands, Belgium, or southern Britain, atmospheric deposition may contribute more than agriculture to the nitrogen that's being added annually.

Kleinjans: The controversy in discussions of the greenhouse effect concerns the temperature increases predicted by the available models. These predict increases of about 2–4 °C in the next century. As I understand it, this cannot be measured against the background, natural or unexplained variations that have been observed during the last 80–100 years. We have to *believe* that there is an enhanced greenhouse effect. One could argue that we are just dealing with the return from the so-called little ice age which occurred during the 16th–17th century.

Oeschger: That's a good point. We need to understand the reasons for the little ice age. It might be caused partly by somewhat lower solar luminosity and partly by a higher dust loading of the atmosphere, but might also reflect an internally introduced change in ocean circulation. This is one of the most important questions which needs to be addressed. During the last 1000 years there have been global temperature changes of up to ±0.5 °C. Since the end of the last glaciation, 10 000 years ago, there have been changes of ±1 °C at most. Thus, during the past 100 years global temperature has still remained within the range of natural variability. We need to use all the information from all the disciplines to disentangle the natural and anthropogenic effects. This will then help us to improve predictions about future climate. However, we have to be aware that the greenhouse forcing has increased significantly. There are uncertainties concerning sensitivity. When and how much will climate change, and will there be unexpected events?

Hautvast: What are the consequences for humans of these environmental changes? Will survival be more difficult? If temperature rises by 1 °C, for example, so what? One could argue that it would be beneficial because we would eat less food, which would help the food problem.

Oeschger: A change of 3–4 °C, which has been predicted, is considerable. The important determinant will be the rate of change. If the change happens relatively slowly, it probably won't be difficult for society to adapt, but if there is a rapid change, 0.3 °C per decade, 3 °C in a century, adaptation will be very difficult.

Zehnder: It has been argued, particularly in the United States, that if temperature increases, vast areas of Canada, for example, would become suitable for the production of wheat. However, the soil there is not in an optimal condition for wheat production; it will take 10 000 years before it is optimal for wheat production. A more important question is how the patterns of rain change, whether dry areas develop, in the United States or Russia, for example.

If the southern part of the USA dries out, will the northern part of North America really be ready for wheat production?

Rabbinge: I think it would be possible to develop presently unused arable land in a short time. The major question here is the potential growth and production of various crops. For example, if temperature increases by 3 °C it will probably become attractive to grow corn in more northerly areas than at present. There can be a shift. We shall come back to this later, particularly when discussing pests and diseases, because not only the crops themselves but also the crop growth-reducing factors may be affected by a change in temperature.

I would think that an increase of 1 or 2 °C will not be a severe problem for humans as such. However, I agree with Professor Oeschger that the rate and scale on which these changes are taking place might be frightening.

Sutherst: Professor Oeschger, you didn't say much about the possible unevenness of heating around the world. What does a 3 °C increase in the average global temperature mean at different latitudes?

Oeschger: The warming will not be symmetrical because the oceans will heat more slowly than the land; the oceans have the ability to mix heat down to greater depths. Warming will be at a minimum around Antarctica. Under the IPCC's business as usual scenario the estimated warming by 2030 for central North America is 2–4 °C in winter and 2–3 °C in summer. Precipitation is predicted to increase in winter and decrease in summer, with soil moisture decreasing in summer by 15–20%. Similar but somewhat smaller changes are predicted for southern Europe.

An important question, difficult to answer, concerns changes at climatic extremes and in extreme events which generally have more impact than changes in the mean climate.

I should stress that the models will never provide precise predictions of what will happen. We know that internal climatic oscillations exist, such as the El Niño southern oscillation phenomenon and the North Atlantic oscillations. These phenomena may be modulated by the changing boundary conditions of the climate system, and new ones may occur. During the last glaciation there were strong changes in deep water formation in the North Atlantic ocean. This raises the question as to whether in a warming earth similar phenomena could be triggered again. The predicted climatic change may not be smooth, and may show unexpected features.

Hoffmann: One thing we have noticed in California is that the wet seasons correspond to an El Niño period. We have substantial yearly rainfall only during an El Niño event. In the interim period without an El Niño cycle we have a drought. This seems to fit quite well with the recent recognition of the cyclic pattern of the El Niño phenomenon, and with our precipitation records in terms of wet years and dry years. We normally have 5–6 dry years between the wet years, and the wet years appear to correspond to El Niño events. There was a small El Niño event this year, and we had substantial precipitation. Before that, we had 6–7 years of near-drought conditions. This may be just an accidental coupling, but there appears to be some correlation.

Lake: Professor Oeschger, you discussed physical feedbacks on temperature, but not feedback effects on the CO$_2$ increase. If everywhere gets warmer, we will burn less fuel to keep ourselves warm. That's a trivial example, but are there major feedback effects which will either make the CO$_2$ increase faster than thought or alternatively cause it to plateau?

Oeschger: There are a great number of feedbacks, including the biological ones, which have not yet been considered in the models. They are difficult to assess. Some information can be obtained from the parallel changes of temperature and greenhouse gases in the past. Examples of possible feedbacks onto the CO$_2$ cycle from climate change include the following: surface water pCO$_2$ will increase with increasing temperature; ocean circulation may change because of decreasing density of the surface water and lower the uptake of excess CO$_2$; the biological pump in the ocean, the rain of dead organic particles, could be influenced by changing ocean dynamics; CO$_2$ fertilization may help to take up excess CO$_2$ in the atmosphere; eutrophication and toxification may change plant growth on a global scale; photosynthesis and respiration tend to increase with increasing temperature; increasing UVB radiation may have detrimental effects on marine and land biota. On the basis of studies comparing the actual CO$_2$ increase and fossil CO$_2$ emission, I have the impression that these feedback effects play a minor role.

Zehnder: We heard earlier that methane produced from rice paddy-fields is an important greenhouse gas. If temperature increases, the areas which are anaerobic will increase. Increasing anaerobic areas will augment methane production, and N$_2$O formation will probably follow. On the other hand, anaerobiosis will prevent the degradation of parts of the organic matter, and oxygen produced during its synthesis will not be reduced. As a consequence, atmospheric oxygen will increase. Are there any scenarios for such a change in concentration of oxygen?

Oeschger: The amount of oxygen in the atmosphere is relatively low, only 20% by volume. However, the CO$_2$ concentration is only 300 p.p.m. The oxygen variations are very small, but they have been observed, and they are an interesting indicator because they reflect, as you say, the whole phenomenon of oxidation.

Lake: Nature has recently published an article by R. F. Keeling, the son of the Keeling who started CO$_2$ measurements, and S.R. Shertz; they are making oxygen measurements with the precision necessary to match the CO$_2$ records. They have already begun to show a pattern of oxygen inverse to the CO$_2$ pattern (Keeling & Shertz 1992).

Avnimelech: Would the increase in dust and suspended particles in the atmosphere increase reflectance and have a compensatory effect?

Oeschger: In the southern hemisphere the temperature increase is much more regular than in the northern hemisphere. Measurements of atmospheric turbidity suggest that the greenhouse forcing in the northern hemisphere is partly cancelled by the effects of dust produced by human activities.

Avnimelech: This suggests a way to stop the increase in the greenhouse effect.

Oeschger: This idea has been put forward many times already. For example, it has been suggested that we should put sulphuric acid into the atmosphere to balance the greenhouse effect. This is not an aesthetically pleasing way to do it!

The excess CO_2 in the atmosphere is integrated whereas the dust is always in a state of equilibrium. With constant use of fossil energy and dust production, the CO_2 would still increase whereas the dust would be constant. This would have only a delaying effect.

Avnimelech: The effect of temperature on evapotranspiration may be highly significant in semi-arid regions. A 1.5 °C increase in temperature would increase evapotranspiration by 10%.

Oeschger: Evapotranspiration would certainly increase. The water cycle and its interaction with plants is an important part of the climatic system.

References

Cirerone RJ, Oremland RS 1988 Biogeochemical aspects of atmospheric methane fluxes. Global Biogeochem Cycles 2:299–327

Keeling RF, Shertz SR 1992 Seasonal and interannual variations in atmospheric oxygen and implications for the global carbon cycle. Nature 358:723–727

Chemical pollution of the environment: past, present and future

Michael R. Hoffmann

Department of Environmental Engineering Science, W. M. Keck Laboratories, California Institute of Technology, Pasadena, California 91125, USA

Abstract.   In an era of 'global environmental change' people are concerned about emissions of CO_2, CH_4, N_2O and chlorofluorocarbons (CFCs) to the atmosphere because of their direct impact on global warming and their stratospheric ozone-depleting effects. Unprecedented efforts have been made to reduce the global emissions of CFCs. Major industries, which are competing within the modern global economy, have recognized the importance of maintaining a 'green' perspective. Future operations will be designed to reduce the direct emissions of chemical by-products to air, water and soil, and to recycle and to reuse critical solvents such as water. 'Star Wars' technologies for the rapid, economical and effective elimination of industrial and domestic wastes will be developed and employed on a large scale. Advanced technologies for the control and monitoring of chemical pollutants on regional and global scales will be developed and implemented. Satellite-based instruments will be able to detect, to quantify, and to monitor a wide range of chemical pollutants. Our understanding of the fate and consequences of chemicals in the environment will increase dramatically such that we shall be able to predict the environmental, ecological and biochemical consequences of novel synthetic molecules with much greater precision.

1993 Environmental change and human health. Wiley, Chichester (Ciba Foundation Symposium 175) p 23–41

In order to understand the future of chemical pollution of the environment—that is, the atmosphere, the hydrosphere, the lithosphere and the biosphere—we must understand clearly the past and the present state of concern in science and engineering about chemical pollutants.

The past

In the latter half of the 19th century concern about the environment was focused on biological and chemical pollutants that had direct impacts on human health. Some of the principal pollutants of concern at this time and in the more recent past are shown in Table 1. During this period, the relationship between human health and environmental contaminants—that is, chemical and biological

TABLE 1   Chemical pollutants of concern in the past

Atmosphere	Water
SO_2	BOD (biological oxygen demand)
H_2SO_4	COD (chemical oxygen demand)
H_2S	TOC (total organic carbon)
Black carbon (soot)	Oil
NaOH	Grease
Pb	Acids
Hg	Bases
CO	Phosphates (PO_4^{3-}, $P_2O_7^{4-}$, $P_3O_{10}^{5-}$)
HCl	Nitrate (NO_3^-) and nitrite (NO_2^-)

contaminants—was first recognized and scientifically established. Physicians noted that the occurrence and frequency of several diseases, cholera, for example, could be correlated with the extent of unsanitary conditions, such as waste water discharges, in the vicinity of drinking water supplies. This phenomenological correlation provided the impetus for the large-scale treatment of municipal drinking water. One of the earliest treatment techniques involved the chemical disinfection of drinking water with chlorine (in the form of hypochlorous acid, HOCl) and later ozone (O_3) was used. The chemical disinfection of drinking water was coupled with the slow filtration of water through beds of sand and gravel in an effort to remove large particles that could harbour infectious organisms. The combined effect of these two approaches to the treatment of drinking water was a dramatic reduction in the rates of transmission of a variety of water-borne diseases such as cholera (which is still a problem in the Third World, as shown by the recent outbreak in Peru). Another major environmental engineering advancement during the 19th century was the introduction of the flush toilet. This device of convenience was combined with underground sanitary sewer systems (earlier sewerage systems were imbedded in the middle of streets) for the rapid collection and eventual disposal of domestic (human) and industrial waste. Waste water flows were collected at central locations where, in some cases, particulate matter was removed by sedimentation. However, in the majority of cases, the collected waste water was discharged directly into receiving waters such as rivers, lakes, estuaries and oceans without treatment. In some cases, waste water was directly applied to land through the use of holding and evaporation ponds. In addition, many industries discharged their chemical waste effluents directly into natural waters.

During the same period there was a growing recognition that the chemical contamination of the atmosphere was having deleterious effects on human health. On several occasions there were dramatic increases in the daily number of deaths in London. These increases in the number of officially recorded deaths for metropolitan London (measured in terms of excess deaths) were correlated with the occurrence of widespread fog, combined with smoke to give smog, cold weather, the excessive burning of high sulphur coal for home heating and industrial energy production, the intense emission of sulphur dioxide (as detected by its peculiar odour) and stagnant air masses (no wind). A figure of 1000 excess deaths per day was not uncommon during major air pollution episodes. When these correlations were established, regulations were introduced sporadically in an attempt to reduce the direct emissions of soot (black smoke), caustic soda (NaOH) and acids (H_2SO_4 and HCl).

Most of the early attempts at control of air pollution were limited in scope and had a negligible impact on overall air quality. The most significant impetus for modern air pollution control legislation and technology was provided by the 'killer fog' of December 1952 in London. During this notorious event, which lasted for five days, more than 4000 excess deaths were recorded, and over the next three months more than 18 000 were registered. This disastrous episode raised the environmental consciousness of the urban, industrialized world to an unprecedented level. Concurrent with the long-delayed recognition of the link between air quality and human health in London was the discovery of another type of smog in Los Angeles. The occurrence of this unique form of air pollution was found to be correlated with the emission of reactive hydrocarbons and nitrogen oxides from motor vehicles, high temperatures (25 °C–45 °C), intense sunlight (high actinic flux), stagnant air (low wind velocity), low temperature inversions (cool air trapped by warmer air above), the daily accumulation of ozone (O_3) and organic peroxides (e.g., CH_3OOH), and the diurnal reduction in visibility (for example, from 16 km down to 200 m) resulting from the formation of a chemical fog. This unusual type of warm weather smog, which appeared to have a basis radically different from that of the traditional cold weather London smog, was named photochemical smog or LA smog. It took more than 25 years for the introduction of effective legislation that brought about concerted efforts to control LA smog.

LA smog is no longer unique to the Los Angeles basin—it has been 'exported' globally. LA-type smogs are now common occurrences in Europe, China, Taiwan, Brazil and elsewhere.

The present

Over the last 20 years the focus of the concerns of advanced technological societies has shifted from local and regional scale activities, such as the treatment of drinking water, waste water, flue gas and exhaust emissions, to global scale environmental problems. These problems include the depletion of stratospheric ozone, global warming, the decrease in the oxidative capacity of the troposphere (that is, the self-cleansing capacity of the atmosphere), the long-distance transport of toxic chemicals such as pesticides, widespread acidification of the atmosphere, global increases in tropospheric ozone, how to clean up ground-water, soils and surface water contaminated with hazardous wastes, how to destroy and eliminate chemical weaponry, the safe storage of nuclear wastes, the contamination of ground-water with chlorinated hydrocarbons and gasoline, bio-mass burning, deforestation and the negative impacts of the green revolution (such as increased release of methane into the atmosphere, increased contamination of ground-water, increased release of nitrous oxide, N_2O, to the stratosphere, and increased application of xenobiotic compounds to soils, water and air).

During the latter half of the 20th century, the basic goals of advanced waste water treatment have been, (1) reduction of the biological oxygen demand (BOD) and (2) lowering of suspended particulate matter concentrations in waste water effluents. The objective was to reduce BOD (as measured over five days at 20 °C) from 300 to 30 mg l^{-1} and particulate matter from 300 to 30 mg l^{-1}, with the aim of protecting natural waters. However, even these simple goals have proven difficult to achieve in practical and economic ways. In some areas, where waste water effluents are discharged directly into lakes, streams, rivers and estuaries, major effort has been devoted to reducing the direct and indirect input of phosphates (PO_4^{3-}, $P_2O_7^{4-}$, $P_3O_{10}^{5-}$) and nitrate (NO_3^-) in an effort to control eutrophication, with moderate success. However, at present, few communities in the world have actually achieved the basic goals of primary waste water treatment and in even fewer places has full-scale secondary waste water treatment been implemented. A very small number of facilities actually use tertiary (advanced) waste water treatment procedures.

In contrast, the goal of providing drinking water which is free of biological pathogens on a large scale has been achieved in the advanced technological countries, although the supply of safe drinking water is still a major goal in the Third World. The major concern today in the advanced countries is provision of water that is also 'free' of undesirable chemical contaminants such as trihalomethanes (e.g., $CHCl_3$), which are formed *in situ* during water treatment through reaction of chlorine with humic substances present in water, and phenolic compounds (such as phenol and catechol) and chlorinated hydrocarbons (e.g., trichloroethylene $Cl_2C{=}CHCl$, pentachlorophenol C_6Cl_5OH, and hexa-chloroethane Cl_3CCCl_3) which are both frequently found in contaminated ground-waters. Another major objective of water treatment is the effective control and elimination of viruses and cysts, which have proved to be more difficult to control than bacteria.

Over the past twenty years, we have shifted our concerns about air quality from the emission of carbon monoxide (CO), sulphur dioxide (SO_2), lead (Pb) and soot (black particulate matter) to nitrogen oxides (NO_x), reactive hydrocarbons (C_7H_8), methane (CH_4), chloroflurocarbons (CFCs, e.g., CF_2Cl_2), carbon dioxide (CO_2), polycyclic aromatic hydrocarbons (PAHs), polychlorinated biphenyls (PCBs) and dioxins (e.g., 2,3,7,8-tetrachlorodibenzodioxin). Sulphur dioxide emissions have been reduced dramatically over the last twenty years simply by changing the particular type of fossil fuel used for combustion. For example, fossil fuels (such as natural gas) with low sulphur content ($<< 1.0\%$ by weight) have been substituted for fuels such as coal and high-sulphur fuel oil which have much higher sulphur contents ($> 3.0\%$ by weight). This strategy has been effective in the Los Angeles basin, where the net sulphur emissions have been reduced from more than 1200 tonnes per day to less than 180 tonnes per day over a period of ten years. As a consequence, the degree of visibility reduction by aerosol sulphate—$(NH_4)_2SO_4$ and NH_4HSO_4, ammonium sulphates in the particulate form—and sulphuric acid (H_2SO_4) in the gas phase and droplet phases has been lessened significantly. The net result of this simple control measure has been a significant reduction in atmospheric acidity and a noticeable improvement in overall visibility in the LA basin.

Since 1974, major efforts have been expended to reduce the emissions of chlorofluorocarbons to the atmosphere because of their clear-cut role in the depletion of stratospheric ozone and in generation of the springtime Antarctic ozone hole. Unprecedented world-wide cooperation has been shown in a concerted effort to reduce or totally eliminate the emission of those CFCs with the greatest ozone-depleting potential. Over the last decade we have come to realize fully that chemical pollution knows no boundaries and that many pollutants are transported over long distances where they have adverse effects that were not anticipated. Thus, chemical pollution control is now a global scale problem. For example, many pesticides such as DDT (dichlorodiphenyltrichloro-ethane) and Mirex have been found in large concentrations in polar regions even though their use in many countries has now been banned. The long-term progressive transport of these semi-volatile xenobiotic compounds into the polar regions via a hopping mechanism from land to air to water has been dubbed the 'cold-finger' effect. In addition to the transport of pesticides toward the polar regions, pollutants have also accumulated in the polar atmosphere in the form of 'Arctic haze'. We now appreciate that, as in the case of long-distance tele-communications, there is no longer a truly remote environment that is totally uncoupled from the impacts (i.e., emissions) of a modern technological society.

An overview of chemical compounds and chemical species of importance in environmental chemistry either as pollutants or as reactive intermediates is given in Table 2.

The future

In the future, major industries, corporations and businesses, which are competing within a global economy, will have to recognize the importance of maintaining a 'green' environmental perspective. Future chemical manu-facturing operations should be designed to reduce the direct emissions of chemical by-products to air, water and soil such that critical solvents such

TABLE 2a   Chemical pollutants of the atmosphere of current concern

Component affected	Type	Species
Global climate	Greenhouse gases Aerosols and clouds	CO_2, O_3, CH_4, N_2O, $CFCl_3$, CF_2Cl_2, H_2O SO_2, $RCHO$, RCO_2H, $(CH_3)_2S$, H_2SO_4, HNO_3, SO_4^{2-}, NO_3^-, RSO_3^-, NH_3, NH_4^+, CH_3SO_3H, $HOCH_2SO_3^-$
Biogeochemical cycles	Carbon	CO, CO_2, CH_4, CH_2O, C_2H_4, C_5H_8, $C_{10}H_{16}$, C_xH_y
	Nitrogen	N_2O, NO, NO_2, NO_3, N_2O_5, HNO_3, HNO_2, $CH_3COO_2NO_2$
	Sulphur	H_2S, $(CH_3)_2S$, $(CH_3)_2S_2$, CH_3SH, CS_2, COS, SO_2, CH_3SO_3H, $HOCH_2SO_3H$
	Ammonia	NH_3, NH_4NO_3, NH_4HSO_4, $(NH_4)_2SO_4$
	Oxidants	H_2O_2, HO_2, OH, O_3, $ROOH$
Stratospheric ozone depletion	Oxygen	O_3, $O(^3P)$, $O(^1D)$, $O_2(^1\Delta_g)$
	Hydrogen	H_2O, H, OH, HO_2, H_2O_2
	Nitrogen	N_2O, NO, NO_2, NO_3, N_2O_5, HNO_3, HNO_2, $HNO_3 \cdot 3H_2O$, $HNO_3 \cdot H_2O$
	Halogens	CH_3Cl, CF_2Cl_2, $CFCl_3$, CCl_4, Cl, HCl, CH_3CCl_3, ClO, $OClO$, ClO_2, $HOCl$, $ClONO_2$, Cl_2O_2, CH_3Br, Br, BrO, HBr, $BrONO_2$
Tropospheric chemistry	Oxygen	O_3, $O(^3P)$, $O(^1D)$, $O_2(^1\Delta_g)$, H_2O_2, H_2SO_5
	Carbon oxides	CO, CO_2, H_2CO_3
	C1 carbon	CH_4, CH_2O, $HCOOH$
	Hydrocarbons	C_2H_4, C_5H_8, $C_{10}H_{16}$, C_7H_8, ROH, C_xH_y, PAHs (polycyclic aromatic hydrocarbons)
	Nitrogen oxides	N_2O, NO, NO_2, NO_3, N_2O_5, HNO_3, HNO_2, HO_2NO_2, NO_y
	Sulphur	H_2S, $(CH_3)_2S$, $(CH_3)_2S_2$, CH_3SH, CS_2, COS, SO_2, CH_3SO_3H, $HOCH_2SO_3H$, H_2SO_5, HSO_4^-, H_2SO_4, $SO_2 \cdot H_2O$, HSO_3^-
	Carbonyls	$HOCH_2(CH_2)_nCHO$, $R(CH_2)_nCOOH$, $RCHO$, $H_2C_2O_4$, $(HCO)_2$, R_1COR_2, R_1COCOR_2
	Metals	Fe, Cr, Mn, Cu, Pb, Pt, Pd, V, Zn, Cd, Ir, Rh, Mg^{2+}, Ca^{2+}, Mn^{2+}, $Fe(III)/Fe(II)$, $Cr(VI)$, Na^+, $Cu(II)/Cu(I)$, Ti, K^+
	Radionuclides	^{222}Ra, ^{85}Kr, ^9Be, ^2H, ^3H, ^{210}Pb, ^{137}Cs, ^{90}Sr
	Isotopes	^{14}C, ^{13}C, ^{18}O, ^{33}S, ^{34}S
	Radicals	OH, HO_2, Cl, ClO, RO_2, RCO_3, NO_3, IO, CH_3, HSO_3, HSO_4, HSO_5, Cl_2^-

TABLE 2b　Chemical pollutants of the hydrosphere of current concern

Type	Species
Organic compounds	Detergents, pesticides, chlorinated phenols, halogenated phenols, DDT (dichlorodiphenyltrichloroethane), DDE (dichlorodiphenylethane), Mirex (hexachloropentadiene dimer), carbamates, chlorinated alkanes, chlorinated alkenes and alkynes, alcohols, chlorinated alcohols, ethers, ketones, amines, nitriles, substituted phenolics, fatty acids, dioxins, alkanes, alkenes, aromatic compounds, PAHs (polycyclic aromatic hydrocarbons), haloforms, humic substances, carboxylic acids, nitrilotriacetic acid (NTA), ethylenediaminetetraacetic acid (EDTA), benzene, chlorinated benzenes, toluene, xylene, gasoline, oil, PCBs (polychlorinated biphenyls), $(CH_3)_2S$, $(CH_3)_2As$, CH_3HgCl
Inorganic compounds	Fe, Mn, Pb, Sn, Co, Zn, Cd, Hg, Pd, Te, Au, Tl, Pt, Cr, Ir, SO_4^{2-}, CO_3^{2-}, HCO_3^-, metal–organic complexes, H_2S, COS, As, SeO_3^{2-}, SeO_4^{2-}, PO_4^{3-}, $P_2O_7^{4-}$, $P_3O_{10}^{5-}$, CN^-, F^-, AsO_4^{3-}, AsO_3^{3-}, NO_3^-, NO_2^-, NH_4^+
Oxidants	O_3, $O_2(^1\Delta_g)$, H_2O_2, OH, HO_2, HOCl, NO_3, HOBr, Fe(III), Co(III), Pt(VI), Cr(VI), Mn(III), Mn(IV), Cl_3CCCl_3, $C_6H_5NO_2$
Reductants	H_2S, N_2H_4, NH_2OH

as water, benzene, carbon tetrachloride, ether and alcohol can be recycled internally.

In many countries commercial concerns are advertising themselves in terms of being friendly to the environment; in Germany, every major company is now doing business in an *umweltfreundlich* way. This approach is not purely altruistic—behind it lies good salesmanship and sound economics. Waste disposal is no longer free; the cost of retroactive cleaning up of hazardous waste is prohibitive. Clean air, soil and water have a definable economic value, which will be reflected in the trading of 'rights to pollute' on the futures markets in Chicago and elsewhere. All countries will come to realize that the environment has a tremendous economic, aesthetic, psychological and cultural value and people will be willing to pay to keep their air, soil and water clean, their fields and forests in a relatively natural state and biological diversity at a maximum.

'Star Wars' technologies for the rapid, economical and effective elimination of industrial and domestic wastes will be developed and employed on a large scale. Advanced technologies for the control and monitoring of chemical pollutants at regional and global levels will be developed and implemented. Satellite-based instruments will be able to detect, to quantify and to monitor a wide range of chemical pollutants. Our understanding of the fate and consequences of chemicals in the environment will increase dramatically such that we shall be able to predict the environmental, ecological and biochemical consequences of novel synthetic molecules with much greater precision.

In the future there will be a renewed concern about the presence of acids in the atmosphere, in clouds and in rain. The problems caused by acid rain (first identified in the 1860s by Robert A. Smith) have not disappeared—they have simply lost the attention of the global community. The problems associated with atmospheric acidification will intensify because there will be a reduction in the sources of ammonia (NH_3), Nature's most common base/acid-neutralizing chemical, in the atmosphere as animal husbandry becomes further confined to limited areas. As the progressive acidification of natural waters continues and as the biological consequences, mainly a reduction of local biological diversity, are realized there will be a renewed call for reduction in the emissions of sulphur dioxide and nitrogen oxides (the precursors of sulphuric acid and nitric acid) into the atmosphere far beyond the limits set by the current clean air laws. The problem of atmospheric acidification will be exacerbated by the increased reliance on coal as a fossil fuel source and a source of basic hydrocarbons for the chemical manufacturing industry. In this regard, advanced cleaning technologies for the elimination from coal of reduced sulphur compounds and trace element contaminants (e.g., As, Se, Cr, Cd, Pb, Hg) will have to be devised. Trace contaminants will have to be removed from coal, concentrated and eventually eliminated or sequestered. Advanced technologies for the reduction of NO and NO_2 (NO_x) to N_2 (such as with NH_3 injection and catalytic reduction) will be developed and utilized on a large scale in all combustion operations. Emissions of hydrocarbons and NO_x from motor vehicles will be reduced because of widespread use of high efficiency diesel-powered engines, through the use of advanced multiple catalytic converters and by a gradual shift to electric and solar-powered vehicles. One can foresee greater exploitation of solar, wind and nuclear energy (provided that safer technologies are developed involving fusion rather than fission) in an attempt to control the chemical pollution of the environment.

In the chemical, drug and manufacturing industries it will be necessary to consider the environmental implications of novel synthetic compounds and composite materials before large-scale commercial production is initiated. The magnitude of the potential problems associated with a particular chemical pollutant will need to be determined before its widespread utilization. Pesticides, which are essential for continuation of the green revolution in agriculture, must be redesigned such that they degrade abiotically or biologically within weeks of application. Utilization of new materials such as NiAs semiconductors must be examined in light of the potential impact of the introduction of large quantities of arsenic into the environment.

The use of alternative chlorofluorocarbons (HCFCs) will be problematic also because they will cause increased production of carbonyl halides (e.g., phosphene, $Cl_2C=O$, and carbonyl fluoride, $F_2C=O$) and trifluoroacetic acid (CF_3COOH) in the troposphere. Trifluoroacetic acid is known to be highly toxic to plants and other organisms, humans included. Because of this problem there will be a growing use of nanocomposites containing optically transparent maghemite (γ-Fe_2O_3) embedded in polymers as magnetocaloric refrigerants for domestic, commercial and industrial cooling (climate control) and refrigeration. Because of the long-standing problems associated with lead and cadmium pollution we shall see a major shift away from these metals as primary constituents of electrochemical batteries.

Alternative chemicals and procedures for the cleaning of electronic devices, metal fabrications and clothes must be developed. Thus, we can foresee a reduction in the discharge of chlorinated hydrocarbons such as methyl chloroform (CH_3CCl_3), trichloroethylene and tetrachloroethylene ($Cl_2C=CCl_2$). For example, one can envisage the application of supercritical fluid extraction (with CO_2 or H_2O) for certain cleaning procedures.

Any consideration of the likely future changes in chemical pollution of the environment must take into account the attitudes and fates of the developing nations and the former communist countries in their transition from socialism to capitalism. Control of chemical pollution may be sacrificed for general economic welfare. However, in light of our knowledge of the fate and transport of pollutants in the environment, the advanced technological countries (the G7 group, at least) must be willing to cooperate with the less fortunate countries to protect global and local environments, to protect human health and to insure the economic well-being of all nations.

Acknowledgements

I am greatful to the Alexander von Humboldt Foundation (Bonn, Germany) for providing the financial support, through the Alexander von Humboldt Prize, that allowed me to write this paper and travel to Europe to attend the associated conference. In addition, I would like to thank Dr Detlef Bahnemann (Institut für Solarenergieforschung, Hannover) and Professor Dr Meinrat Andreae (Max-Planck-Institut für Chemie, Mainz) and Professor Dr Rudi van Eldik (Institut für Anorganische Chemie, Witten-Herdecke) for their support and encouragement during my stay in Germany.

DISCUSSION

Lincoln: London smog is really a thing of the past. We are now faced with a new environmental problem, the modern-day smogs in the major cities of the South, Mexico City, for example. Is the chemistry of this like that of the Los Angeles smog?

Hoffmann: Yes. People would often say to environmental chemists who were studying the environment in Los Angeles that LA smog was an anomaly, that this type of chemistry did not occur anywhere else. We now realize that it occurs everywhere, in The Netherlands, in Germany, even in Switzerland. The problem in Mexico City, for example, is essentially related to high light intensity because of the city's high altitude and stronger inversion layers because of its valley location, coupled with stagnant high pressure systems that form over the Mexico City area. With the valley inversion, temperature change varies as a function of altitude, which prevents vertical mixing. The nitrogen oxides directly emitted react with hydrocarbons. The basic chemistry takes place in Tokyo, Mexico City, everywhere. This is the new smog, or the old LA smog.

The real question from a public health stand-point is, what level should we set as the standard for air quality? In California the standard for ozone is about 100 parts per billion by volume, and laboratory studies are showing that there are effects on health at half that level, certainly in well-conditioned athletes and asthmatics. There has been much debate in the USA as to whether if that standard is lowered industry will be able to afford it. Los Angeles is the only city that frequently breaks that particular standard.

Kleinjans: I was puzzled by your remarks on newly introduced chemicals. One would expect that guide-lines such as those of the FDA (Food and Drug Administration) on toxicity screening would ensure that these chemicals are safe as far as we can tell.

Hoffmann: I am sure that some of the dangerous chemicals will be caught by toxicity studies, but I am not so sure about those from large-scale industrial products or materials that have large-scale applicability.

Klein: They all have to go through testing procedures.

Hoffmann: But for what effects are they tested? Are all the chemicals tested against asthmatics? Are *Salmonella* mutants the proper test organism for mutagenicity in all cases?

Lake: I would like to ask about 'natural' background levels of some of the chemicals in the human environment that you have identified as being toxic. EERO has just made an assessment of dioxins and health (Ahlborg et al 1992). Bacteria exist that can use dioxins as a substrate to produce energy, which suggests that there are naturally occurring dioxin-like chemicals as well as those mankind has added. Could you say something about the natural background levels of some of the chemicals that you have discussed? We shouldn't get too scared about the current levels until we know the background levels that may have been around since the middle ages.

Hoffmann: I doubt whether dioxin was present in the middle ages as dioxin. There were certainly polycyclic aromatic hydrocarbons, produced by high temperature combustion; one can trace these back in sedimentary environments. Levels of DDT, PCBs and dioxin in lake sediments are essentially zero before their commercial introduction. The time at which they appear in accumulated sediments corresponds to their introduction in industry. In the case of dioxin, we now realize that 2,3,7,8-TCDD (tetrachlorodibenzo-*p*-dioxin), which is the toxic compound of major concern, is produced in high temperature combustion processes, because of the contamination of various plastic products with things like pentachlorophenol. Pentachlorophenol fuses at high temperatures in a relatively simple reaction that leads to the production of dioxin. Dioxin profiles in sediments are not found before the synthesis and use of pentachlorophenol. Dioxin is a by-product of pentachlorophenol production, and of production of the herbicides 2,4-D and 2,4,5-T.

Zehnder: As far as we know, 2,3,7,8-TCDD is not degraded by biological mechanisms. Non-chlorinated, monochlorinated and dichlorinated dibenzo-*p*-dioxins are metabolized by microbes in the presence of oxygen. There is evidence that octachloro- and 1,2,3,4-tetrachlorodibenzo-*p*-dioxins are reductively dehalogenated to monochlorodibenzo-*p*-dioxin in sediments under anaerobic conditions. Our present knowledge of the biodegradation of dioxins is certainly not complete. I am sure that we shall learn more about the metabolic potential of microbes towards dioxins, just as we have with other compounds. When thinking about the biodegradation of dioxins, we have to bear in mind that all these compounds are very hydrophobic. They are adsorbed onto surfaces and are not generally available to microorganisms. So, in many cases, it is not the metabolic potential of the microbes that is limiting but the physical properties of the compounds which reduce their bioavailability.

Hoffmann: The natural background concentration of ozone varies from about 20 to 30 p.p.b. by volume on a global basis, in comparison with a pollution threshold of 100 p.p.b. Some people, especially highly conditioned athletes, suffer from exposure to ozone at 50 p.p.b., which is only a slight increase above what we consider to be normal background levels. What is 'normal'? As long as there are reactive hydrocarbons emitted into the atmosphere that participate in the chemical reactions that lead to photochemical smog, I don't know how to define 'normal'.

Klein: In my institute studies are being done on East German cities such as Bitterfeld investigating the effects of SO_2 concentration and variations in this on the health of human populations. So far, no clear-cut relationship has been found between SO_2 and human health; other factors such as eating and drinking behaviour may mask these effects tremendously.

The concentration of SO_2 gas is not correlated as it is in West Germany with the acidity of the aerosols, SO_2 adsorbed or chemically bound to solid particles. This is important because inhalation of aerosols is a major factor in respiratory disease.

Hoffmann: This is a complex problem. I took the London fogs and SO_2 as an illustration. It's difficult in laboratory studies with humans to really probe the effects of exposure to concentrations that would be experienced during extreme events. There are some environmental exposure chambers at the University of California at San Francisco, but the problem is that the people who are most sensitive to air pollution can't really be used as test subjects.

You need someone who is willing to go into an exposure chamber, and the levels are always maintained at a concentration below the reported toxic threshold. It's difficult to get actual reproducible laboratory evidence about the effects of human exposure to trace atmospheric components.

Klein: There are some extremely valuable field studies underway combining epidemiology with chemical analysis and with medical analysis of the patients, finding out the whole picture.

Another interesting example is that of childhood leukaemia. It has been argued that the incidence of childhood leukaemia is increased close to nuclear power stations. The increased probability of 5 in 100 000 is small but potentially significant. An epidemiological study all over Germany has revealed that higher than average rates are found not only in places where nuclear power stations exist, but also in places where nuclear power stations are *planned* (Michaelis 1992). There are many factors which we don't yet understand.

Of course, we like to discuss the relationship between air pollution and health, because we can measure air pollution much more easily than soil pollution or water pollution. It's clear to me that the major impact on health comes from the chemicals in the food chain. Chemicals accumulated through eating and drinking contribute more than those acquired through breathing. This should be considered in any discussion of the effects of pollution on human health. My feeling is that, in the long run, it is the cleanliness or pollution of water resources which will be more critical to human health, because the water compartments of the environment are accumulating chemicals all the time. Although the buffering capacity of ground-water reserves is high, reserves of high purity will become reduced and this will be the real future problem.

Hoffmann: I am not discounting water as a major problem—I actually work in both compartments—but the health impacts in relation to concentration are more clear-cut with air than with water.

Hautvast: One point we should consider in this discussion about the toxic effects of chemicals is the immune status of the human body. Consider a person in a developing country with a poor immune status because of deficiencies in micronutrients such as vitamin A. Can we assume that the effect of a toxin on such a person will be greater than on a Western person? This issue is often neglected, although in recent years the role of vitamin A and other micronutrients in immune status has become more evident.

The second point I would like to make concerns the relative importance of air and food pollution. Doll & Peto (1981) have shown in epidemiological studies on toxic substances and cancer that perhaps 1% of cancers can be ascribed to toxic chemical substances. This is still fairly important.

Hoffmann: Bruce Ames, who has done a lot of work in this area, believes that the majority of chemical carcinogens entering the body do so in food. For example, in his mutagenicity studies, he shows a strong correlation between aflatoxin present in peanut butter and the mutagenicity of peanut butter, which is a staple of the USA diet. Peanut butter contains a relatively high concentration of aflatoxin. If his results could be used to predict effects, the prediction would be that everyone in the USA should be dead because of the amount of peanut butter they have eaten over their lifetimes.

Hautvast: The body can cope with many things; we still don't appreciate the strength of the human body.

Kleinjans: Whether or not food is the main route of intake depends on the environmental fate of the chemical in question. For benzene or radon, food is not the main route of intake. I would agree that, in general, the drinking water supply will be one of the most critical issues in the near future; in Western countries we are at the very beginning of a serious problem in relation to nitrates and pesticides in surface water as well as ground-water.

Zehnder: It is technically feasible to remove contaminants from water at the site where it is consumed. Air contaminants can be removed only by wearing a gas mask. In developed countries we are now able to produce good quality drinking water from almost any liquid that contains water, provided that the financial resources are sufficient. From a consumer's point of view, air and water could be cleaned where they are polluted, but only water can be cleaned at the place of its consumption without hindering an individual's lifestyle.

Klein: There are technologies for cleaning the air. We don't simply have to live with the air which is around. There could be state control to clean air at the site of pollution.

Rabbinge: We are discussing micro-elements and trace gases and not the macro-pollutants such as ammonia and SO_2. The concentration of SO_2, for example, in the air has decreased considerably during the last few years, particularly in Western Europe, because it is quite easy to eliminate at the site of production (electrical power stations, mainly). The situation is not the same with many of the trace elements. I would be interested to hear if there is a danger from micro-elements associated with climatic conditions. Some are saying that food is more important, others that air is more important. What are the real dangers to human health of these trace gases? Are there, for example, synergistic effects between them?

Elliott: It's true that people living in an urban environment are exposed to higher levels of benzene, lead, etc. than those in a rural environment. The difficulty is in going from these observations to establish any aetiological link, because many other factors are different in urban environments. Smoking, for example, which hasn't so far been mentioned, is a major cause of morbidity and mortality. Most epidemiological studies concentrate on individuals. We know that there are large individual differences in risk, related to smoking and diet in particular. It's also true to say that we don't know the effects on human health of many chemicals in the environment. New chemicals may go through all sorts of screening procedures, but with combustion processes, for example, we don't know what we are actually exposed to in the vicinity of a plant, and there are many combinations of compounds that have not been tested. The ultimate test is whether there is an effect, direct or indirect, on human populations, although this may be impossible to determine with any certainty. The difficulty is to answer this question against a background of large individual differences in risk related to lifestyle, social class, diet and smoking. The effect of a chemical may be extremely important, increasing disease risks by, say, 10 or 20%, but this might get lost in the background. To study these effects we need much better measurements of pollution, and much better measurements of the health effects. In some countries, including the UK, we are now able to get much better data on health at the local level, for small areas, and we need better measurements of pollution at comparable geographical resolution, to look at local relationships between pollution and health. For example, we don't know the effects of dioxins on human populations. There have been studies of occupational exposure, but we really do not know what goes on outside chemical and combustion process plants, although studies following the explosion at Seveso so far have been reassuring. In the UK, we are carrying out a study of health effects around municipal incinerator plants.

Avnimelech: Public opinion is one issue we haven't considered. I haven't heard people complaining about peanut butter or micropollutants in water, but people do complain about air quality. As far as the well-being of the individual goes, air pollution is very important.

Mansfield: I'm a little uncomfortable about discussing pollution of air and water and food as separate compartments; I think we should be looking at the world as a single compartment, because there are so many transfers of substances and derivatives across boundaries. Some pollutants, ozone for example, are essentially confined to one compartment, but they are very few. Is there enough research on transfers between compartments and the controlling processes?

Hoffmann: There has been a fair amount of active research by environmental chemists interested in multiphasic transfers between air and water, between air and particles, and subsequent deposition of the particles into water and onto land. The use of polychlorinated biphenyls (PCBs) has been banned in the USA since about 1972. The net influx of PCBs into the USA has decreased dramatically, but we are finding that PCBs are coming out of the water into the atmosphere now that the atmospheric concentration is lower than that in water. Chlorinated hydrocarbons, DDT, Mirex, PCBs, and even dioxins are slowly migrating to the polar regions through a hopping mechanism from land to air to water. Because of their physicochemical properties these compounds move to cold regions. In chemistry, when we want to isolate a volatile component, we put a cold finger on a gas vacuum line and we freeze out the component with methanol and dry ice or liquid nitrogen. This same 'cold finger' effect is taking place in the environment. Very high concentrations of Mirex, a pesticide used primarily in the southern part of the United States, are found in Arctic snow. The chemical components which transfer between phases are those which aren't degraded biologically in soil or water, so are transferred back into the atmosphere, adsorbed onto particles and deposited, and eventually migrate towards the poles.

Avnimelech: Is there transfer to vegetation?

Hoffmann: Vegetation takes up some chemicals, but the compounds we are concerned with are semi-volatile, so unless they're biologically attached, as dioxin is to a specific receptor, they partition between the gas and solid phases. The primary thermodynamic driving force is to the atmosphere.

Elliott: The issue of human exposure to dioxin mainly concerns food chain exposure and not local air pollution.

Klein: We consider that the key problem is not the physical chemistry of transfer from soil to water but transfer from the soil to the root or the leaf of a plant, followed by transformation within the plant and passage along the food chain to humans. In this transfer from the geosphere to the biosphere there are many problems to be investigated and solved before we can assess the risks of these compounds to humans.

Edwards: There is a distinction to be made between private air and public air. Doll and others have clearly shown that pollution of private air—smoking—is a major contribution to various disorders, carcinomas and bronchitis in particular. One needs to distinguish between these two because their control is quite different.

There had been some rather remarkable figures which were claimed to show that most cancers were related to industrial processes, and administrative and other actions in the USA were threatened. Doll & Peto's work (1981) clearly put this into perspective, but the methods they used necessarily excluded interactions, because the sum of all the causes was 100%. If interactions are important, which seems likely, then the figures Doll & Peto gave must be regarded as too low.

James: This is a fair comment. Work being done, in The Netherlands in particular, on toxicology and its interaction with nutrition is highlighting the importance of the amplification and synergy that there can be between a toxicant's metabolic effect and its modification or even induction by diet.

Elliott: Doll & Peto's analysis did suggest that, for example, PCBs may be responsible for perhaps a 10% increase in lung cancer, but this is indistinguishable against the effects of smoking (Doll & Peto 1981). Ten per cent of a common cancer like lung cancer is a lot of cases. The problem is to quantify these risks against the large variations in individual risk related to individual air pollution from smoking.

Kleinjans: I think we have to accept that hypothesizing about synergism goes beyond the scope of our thinking. Professor Vic Feron, a well-known toxicologist at CIVO-TNO Zeist in The Netherlands, raised the question: which is more harmful to human health, a glass of milk with a high fat level, or a glass of milk with a low fat level and a trace of dioxin? This puts risks into perspective.

Zehnder: We are shifting away from the central theme of this symposium, which is environmental change and human health. We are looking backwards, trying to make some forward interpretations. Could we now try to combine the two talks we have just heard? If temperature increases globally by 4 °C, it might increase locally by 10 °C or more; an increase of 10 °C would actually double the rates of physical processes. What would this mean for atmospheric processes?

Hoffmann: Some people think a lot more CO_2 and methane will go into the atmosphere than at lower temperatures. Rates of production of methane and nitrous oxide, for example, will increase. Many of the reactions are temperature sensitive, and those will increase to some extent. Chemical transformations in the atmosphere in general will increase.

Avnimelech: Is this good or bad?

Hoffmann: With respect to ozone production in the troposphere it's probably not good.

Zehnder: Some of the hydrocarbons you mentioned, such as methane, are a source of water in the stratosphere. What is the a short-term effect of injection of water into the stratosphere? Will it affect ozone production? What will be the long-term effects?

Hoffmann: One of the major global scale concerns focuses on the lowering of the oxidizing potential of the atmosphere. The oxidizing potential of the atmosphere is determined by the steady-state concentration of OH radicals, which is extremely low. OH radicals are present in the atmosphere at a concentration of less than a part per trillion, at about $10^5-10^6/cm^3$. As more and more pollutants are discharged to the atmosphere and compete for OH radicals, many more species will be allowed to diffuse to the stratosphere where ultraviolet light is more intense, and there will be more potential chemical reactions and thus distortions of the stratosphere. Within the atmospheric chemistry community, this depletion of oxidizing potential is a major concern.

Klein: In trying to combine these two issues, the aspects of global warming as a physical problem and pollution as a chemical problem, one has to distinguish between those compounds which may become distributed all over the world and which will then contribute in some way to global physical effects with an impact on the biosphere and plant growth and in this way affect human society, and those compounds which are more *locally* distributed and can be handled by on-site prevention or remediation technologies.

One thing we haven't discussed is what global warming means in terms of regional warming. In one way we have to think more globally, but in asking what the impacts on human beings will be we need to think on a more regional level. An average figure of 1 °C doesn't tell us anything. Here we have conflicting arguments and real scientific problems in the correlation of global, regional and local phenomena.

Zehnder: Can we then conclude that until now this relationship has not been addressed sufficiently? We have discussed water, drought and agriculture, and air pollution, soil pollution and water pollution, and the effects of all these factors on humans. There might be negative synergistic effects. Global warming might increase certain reactions, such as volatilization of chemicals, and their transformation or incorporation into plants, and into the food chain. Is there any quantitative research being done on the impact of global warming on chemical and biological processes?

Oeschger: The impacts of global warming on chemistry and biology are important but are difficult to address in a quantitative sense. This is why we are also interested in the history of atmospheric components. From the ice records, the general tendency is that with an increase in temperature, levels of trace gases go up. There will be biotic feedbacks onto a warming climate system which are not yet considered in the model predictions.

Sutherst: The interests of developing countries are rather underrepresented here. When I go to developing countries I see the deteriorating conditions there, and it seems to me that pollution is often involved. Making 10% of cars in California electric will have absolutely no effect in comparison with the effect that will result when every Chinese household has a refrigerator. I am also concerned about claims that it's expensive to reduce energy consumption. Amory Lovins' (1989) research in the USA shows that energy conservation can be used to reduce electricity consumption by 75% at a much lower cost than just running a coal or nuclear power station. The technology is already available. On the one hand, human society responds well to local situations such as the smog in London, but, on the other hand, we don't seem to be so good at addressing major, global problems. At the global level we seem to be immobilized by politics. Greenpeace has now targeted chlorine as the major pollutant in the world; it is the source of most of the polluting chemicals in the world. I wonder how much more effective these sorts of non-governmental responses to global problems can be than our present political efforts. What responses can we expect on a global scale in developing countries as well as in affluent countries in the future?

Hoffmann: Public opinion certainly plays a large role in determining what is actually accomplished. I have been working in a variety of different environmental research projects over the past 25 years and have seen interests come and go. There's a great deal of interest for a short period then it dies away. Public and political opinion rises in concern over a particular problem, the research increases, we learn more about the problem, but then people get bored with the problem, even though it still exists, and their interest shifts to a new topic. This happened with atmospheric acidity. There was a flurry of activity in the USA and in Europe that lasted for about 10–15 years. Now, it's virtually impossible to get continued funding for research on atmospheric acidity. The problem hasn't gone away, but the public interest has. This situation also arose initially with CFCs. When Molina & Rowland (1974) first proposed the theory of stratospheric ozone depletion there was much activity in the USA, and there was tremendous dispute between the major producer of CFCs, Du Pont, and the scientific community about the relative role of CFCs. The problem more or less died away until the discovery of the Antarctic ozone hole rejuvenated interest in the phenomenon of stratospheric ozone depletion. If you are doing research on global climate change, you will get funding in the United States; other areas, which may be equally important, are not being focused on at the moment. Public opinion and political opinion are major driving forces.

Mansfield: This point about fashions in research is an important one. You mentioned acid rain, but another example is ozone and crop production. There are now reasonable models for predicting the effects of ozone on soya bean production in the USA, so we think we know all about ozone and legume production. In fact, we don't. We can't advise India, for example, about the likely effects of an increase in ozone on their rather special leguminous crops, of which there are many. At present, there seems to be no way to achieve, globally, coordination in that sort of sustained research, or funding of research which is of international rather than just national interest.

Zehnder: We have talked about air pollution and its effect on human health, which is largely a toxicological and medical problem. We generate air pollution every day and politically we are unable to drastically reduce or ban unnecessary pollution. What happens in the Third World? We haven't yet answered this question. It's somehow too far away from us, and there is seemingly no big issue like the ozone hole over Antarctica to shake people up. From time to time people are shaken up when hungry people are shown on their televisions, but the concern does not last very long. The world is not only Europe and North America and Russia—the world is also Africa, India, and Central and South America. In the future, these will be the world and we, Europe and North America, will become small in comparison. It's important to see all the aspects we have discussed in relation to the development of the Third World. Within 50 years, these continents will be the most important and influential in the future of our globe. Very probably, it will be in these countries where global warming and pollution will have their main effects, not only on public health but also on the development of the various societies.

References

Ahlborg UG, Brouwer A, Fingerhut MA et al 1992 Impact of polychlorinated dibenzo-*p*-dioxins, dibenzofurans and biphenyls on human and environmental health, with special emphasis on application of the toxic equivalency factor concept. Eur J Pharmacol 228:179–200

Doll R, Peto R 1981 Causes of cancer. Quantitiative estimates of avoidable risks of cancer in the U.S. today. Oxford University Press, New York

Lovins AB 1989 Energy, people and industrialization. (Hoover Inst Conf Hum Demogr Nat Resour, California, 1–3 February 1989) Hoover Institution, Stanford University, Stanford, CA

Michaelis J 1992 Untersuchung der Häufigkeit von Krebserkrankungen im Kindesalter in der Umgebung westdeutscher kerntechnischer Anlagen 1980–1990. Institut für Medizinische Statistik und Dokumentation der Johannes Gutenberg-Universität Mainz publication series, Mainz, Germany

Molina MJ, Rowland FS 1974 Stratospheric sink for chlorofluoromethanes: chlorine atom-catalysed destruction of ozone. Nature 249:810–812

River Rhine: from sewer to the spring of life

Alexander J. B. Zehnder

Swiss Federal Institute of Technology (ETH) Zürich, Federal Institute for Water Resources and Water Pollution Control (EAWAG), Ueberlandstrasse 133, CH-8600 Dübendorf, Switzerland

Abstract. Water is the key issue in a number of declarations made by several eminent international commissions in recent years. The availability of clean and unpolluted water is crucial to sustainable development. The River Rhine was turned by pollution into the sewer of Western Europe; environmental protection measures, changes in industrial production and consumers' behaviour and remedial measures have drastically improved the quality of Rhine water, which, besides being the main water-way in Europe, also serves as a source of drinking water for a large population and is used for recreational purposes. Small occasional accidents, major spills, very remote accidents, war activities, etc. threaten the full recovery of what is the socially and economically most important watercourse in Europe. The organizational and technological measures taken to protect the Rhine from pollution can serve as an example of how other major freshwater sources could be protected from contamination or how existing pollution could be remedied.

1993 Environmental change and human health. Wiley, Chichester (Ciba Foundation Symposium 175) p 42–61

Water is a key issue in all recent international declarations on the future (sustainable) development of human society and our globe. In the third principle of its Tokyo Declaration, the World Commission on Environment and Development (WCED) (Brundtland Commission) states that sustainability requires the conservation of environmental resources such as water, and water should be used efficiently (WCED 1987). ASCEND 21 (Agenda of Science for Environment and Development into the 21st Century) recommends research on and studies of hydrological cycles at local and regional scales. In the Dublin Statement of the International Conference on Water and Environment: Development Issues for the 21st Century (ICWE), we can read 'Scarcity and misuse of freshwater pose a serious and growing threat to sustainable development and protection of the environment. Human health and welfare, food security, industrial development and the ecosystems on which they depend, are all at risk, unless water and land resources are managed more effectively than they have been in the past' (International Council of Scientific

Unions [ICSU] 1992). Good quality water is one of the pillars of sustainable development. Only the prevention of its misuse and pollution will allow mankind to survive in the future.

Availability and use of water

The availability of water is determined by the hydrological features of water sources and their boundaries, which are naturally confined to drainage basins or ground-water aquifers. The importance of water resources depends on the needs of the society. Interestingly, cultural developments have been stimulated in large river basins such as the Nile, Euphrates/Tigris, Ganges and the Yellow River. Lack of water may stop regional developments, as in the Sahel zone. Naturally, availability alone is only an indicator of whether abundance or scarcity is more likely. Not all water made available by Nature can be used, and the actual withdrawals depend upon the needs of people living upstream or downstream of a community. Use of water by different sectors may differ considerably, according to the status of industrialization and agricultural practices (Table 1). Globally, four categories of countries can be defined (World Resources Institute [WRI] 1986): those with very low availability of water per capita (1000 m^3 per year or less), those with low availability (1000–5000 m^3 per year), those with medium availability (5000–10 000 m^3 per year) and those with high availability (10 000 m^3 per year or more); the percentage of countries falling into each of these categories is, respectively, 14%, 37%, 14% and 35%.

TABLE 1   Use of water by different sectors in selected countries

Country	Use (%) Public supply	Industry (processing)	Power (cooling)	Agriculture (irrigation)
Sudan	2	0	0	98
Togo	90	0	0	10
Algeria	13	6	0	81
Argentina	9	8	10	73
USA	10	11	38	41
India	3	1	3	93
Indonesia	95	5	0	0
Japan	17	33	0	50
New Zealand	52	11	23	14
Germany	10	35	55	0
The Netherlands	4	24	40	32

Data taken from Meybeck et al 1989.

Pollution of water

Over the past decades, the quality of natural water has deteriorated through the impact of various human activities. The recognition of a pollution problem usually takes considerable time, and the application of the necessary control measures takes even longer. Medieval reports and complaints about inadequate disposal of excreta and of foul and stinking water in overcrowded cities were an early manifestation of water pollution. In 1540, for example, Emperor Charles V left Amsterdam on the day he arrived, moving his headquarters to Haarlem, because he found the quality of the drinking water inadequate. When Emperor Napoleon I visited Amsterdam in 1811, he found the same low quality drinking water and he gave the order that work to improve the quality of the water should be undertaken immediately. The visits of the two emperors to Amsterdam are nice examples of how problems with water quality can be dealt with. Charles V recognized the problem but left the city with its polluted water, turning to nearby alternatives for himself and his court; Napoleon I recognized the problem too, but, instead of abandoning the city, instigated measures to improve the situation.

The first time that a clear causal linkage between water quality and human health was established was in 1854, when John Snow traced the outbreak of cholera epidemics in London back to the Thames river water which was severely polluted with raw sewage (Meybeck et al 1989). Consequently, techniques were developed to treat water before its consumption. Towards the end of the 19th century it was recognized that the source of pollution—the sewage—has to be treated as well. The first processes were anaerobic, then later aerobic procedures were also applied. The aerobic activated sludge process was invented in 1914 by Ardern & Lockett (1914a,b). Both treatment approaches, aerobic and anaerobic, were designed to reduce the overloading of surface water with easily biodegradable organic material. This overloading caused severe seasonal depletion of oxygen. In the 1960s it became apparent that removal of organic matter alone was not sufficient to keep a high water quality standard and to prevent formation of excess algal blooms (eutrophication).

Decaying algae cause depletion of the dissolved oxygen in the water. Oxygen depletion causes an array of problems, including the death of fish and liberation of corrosive and toxic gases. Eutrophication can be combated only by the reduction of essential nutrients, phosphate in particular. Heavy metals and organic micropollutants (compounds at low concentrations that can have large polluting effects) became the major issue in the late 1970s and early 1980s. Treatment processes could not remove these pollutants entirely, and it became necessary to prevent them from getting into water by reducing their release at source. If we are to achieve sustainability, besides reducing pollution, we need to conserve and remediate bodies of water such as lakes, rivers and ground-water, to allow them to return to a state in which they can serve as safe water resources for humans, animals and plants, while still fulfilling their ecological roles.

Using the River Rhine as an example, I shall discuss how the perception of water quality has changed over the last two decades. The Rhine has been used as sewer for one of the most heavily populated and industrialized areas in Europe, but it also served and still serves as a major source of drinking water for more than 20 million people. The remedial measures taken are showing their first results now. The Rhine probably provides the best example of how many other rivers and bodies of water should be treated and managed in the future.

The River Rhine

Among European rivers the Rhine is second only to the Danube in terms of length (1320 km), drainage basin (252 000 km^2, Fig. 1), and mean annual flow (1028 m^3/s in Basle and 2330 m^3/s in Lobith at the Dutch–German border). However, it is the most at risk of pollution, because it is the most important navigable artery in Western Europe. Almost 40 million people live in the Rhine basin and a large proportion of European industry and almost 20% of the world's chemical industries are also concentrated here.

Pollution by heavy metals: cadmium and lead

Cadmium and lead are taken as examples of heavy metal pollutants because they have no biological role, are ubiquitous, ecologically a nuisance and in addition are hazardous. Cadmium is mainly used for electroplating, and in paint pigments, batteries (e.g. nickel–cadmium batteries), plastic stabilizers and alloys. World-wide interest in cadmium was aroused when itai-itai disease in Japan was reported to be associated with the element. The use of cadmium therefore declined drastically, first in Japan between 1969 and 1970, and later in the USA and Western Europe when it was found that cadmium raises blood pressure in animals (Nriagu 1980, 1981).

The use of lead has increased with the growth of the automobile industry; the use of lead for storage batteries and alkyllead fuel additives by this industry represents about 50% of total lead consumption. Cable sheathing and pigments take another 25% of the lead produced. The fuel additives are mainly responsible for diffuse lead pollution of the environment (World Health Organization [WHO] 1977). Mankind has been exposed to lead for millennia. Egyptian women used facial cosmetics containing lead, and the Romans drank wine from lead goblets. The effects of lead poisoning were described by Hippocrates and Paracelsus, but it was 1831 before lead was identified as the causative agent of these poisonings.

From the mid 1970s, the cadmium concentration of Rhine water began to decrease gradually and reached very low levels by 1985. These 'background levels' have remained low ever since (International Commission for the Protection of the Rhine against Pollution [hereafter referred to as the

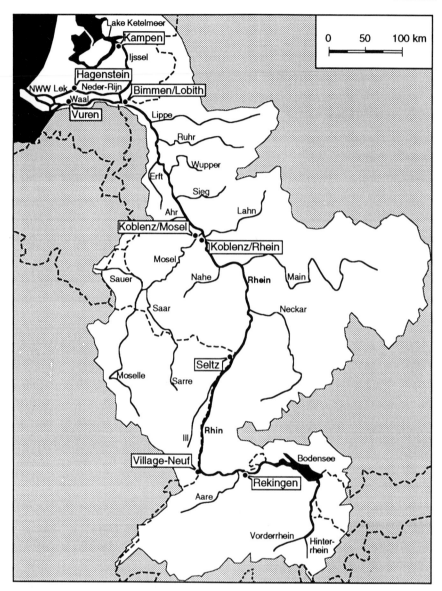

FIG. 1.   The drainage basin of the Rhine, with the major observation points labelled.

International Rhine Commission, IRC] 1988; Fig. 2A). Cadmium levels in the sediments of Lake Ketelmeer reached a peak in 1960 and started to decline only after 1980 (Beurskens et al 1993; Fig. 2B). The northern branch of the Rhine, the IJssel River, which splits from the main stream after Lobith, flows into

47

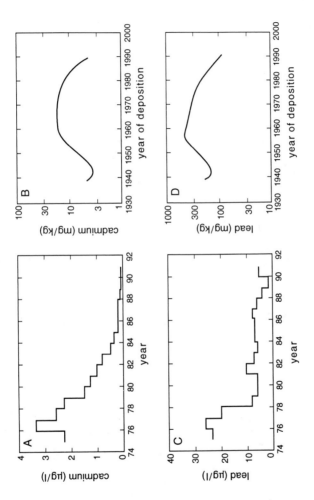

FIG. 2. Concentration of total cadmium (A) and total lead (C) in Rhine water measured at Lobith and of total cadmium (B) and total lead (D) in sediments of Lake Ketelmeer. Data for the concentrations in Rhine water are from International Rhine Commission (IRC 1988) and the Statistical Tables for 1990 (IRC 1990a). Lake Ketelmeer sediment data are according to Beurskens et al (1993). Note that the vertical scales in B and D are non-linear.

Lake Ketelmeer. The geological history of this lake is well documented. It was created as the indispensable outlet for the IJssel water in between two polders. Since its completion in the early 1950s, the lake has acted as a major sedimentation area for the IJssel water. There is no source of pollution along the IJssel River, so Lake Ketelmeer sediments are suitable for at least the qualitative study of pollution of the Rhine over the past 50 years.

Lead concentrations in Rhine water also declined after 1975 (Fig. 2C) but a rather high background level remained after 1980. Nevertheless, a slight decrease from about 1985 can be seen. Most of the lead in the Rhine originates from privately owned motor vehicles. Only the stepwise replacement of leaded by unleaded petrol will substantially reduce the lead concentration. There has been a five-fold reduction in lead concentration in the sediments of Lake Ketelmeer since 1955 (Fig. 2D).

Pollution by organic micropollutants: polychlorinated dibenzo-p-dioxins (PCDDs) and polychlorinated biphenyls (PCBs)

PCDDs and PCBs can be found ubiquitously. They were released into the environment until the early 1970s, when their emissions declined drastically after the accident at Seveso, Italy (PCDD), and after an unforeseen world-wide environmental contamination was discovered. The dioxin problem was first observed in connection with teratogenic effects of the herbicide orange, (2,4,5-trichlorophenoxy)acetic acid (2,4,5-T). These effects were caused by 2,3,7,8-TCDD (2,3,7,8-tetrachlorodibenzo-p-dioxin) present as a major contaminant in 2,4,5-trichlorophenate, which is used for the manufacture of 2,4,5-T. Significant amounts of 2,4,5-T were synthesized world-wide specifically to produce Agent Orange, a mixture of the n-butyl esters of 2,4-D, (2,4-dichlorophenoxy)acetic acid, and 2,4,5-T, which was sprayed as a defoliant in vast amounts over tropical forests and crop-growing land by the US Army during the Vietnam war. PCDDs are also contaminants of chlorophenols and are found in fly ash, produced by incineration of the pulverized fuel burned in power stations (Rappe 1980).

PCBs have been produced commercially since at least 1930, but their production has now been stopped. They were and still are used in industry as heat transfer fluids, hydraulic fluids, solvent extenders, flame retardants, organic diluents and dielectric fluids. It is estimated that of the one million tonnes of PCBs which have been produced, more than one third has entered the environment. Although occupational exposure to PCBs may have adverse health effects, environmental uptake is unlikely to cause such effects (Safe 1987).

PCDDs have not been routinely measured in the Rhine, but the sediments from Lake Ketelmeer have been tested; these reveal a relatively high level of pollution by 2,3,7,8-TCDD between 1950 and 1975, with the peak at 1965 (Fig. 3A). This high level might be related to the production of 2,4,5-T by a German

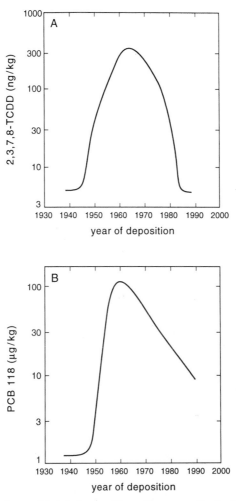

FIG. 3.   Concentrations of 2,3,7,8-tetrachlorodibenzo-*p*-dioxin (2,3,7,8-TCDD) and
2,3′,4,4′,5-pentachlorobiphenyl (PCB 118) deposited in the sediments of Lake Ketelmeer
over the last 50 years. Data are from Beurskens et al (1993).

chemical manufacturing facility located on the Rhine. It is likely that much of
the 2,4,5-T produced was exported to the USA for use in Vietnam. Production
capacity at this plant was 1000 tonnes of 2,4,5-T in 1980. Production was stopped
in 1983 (Beurskens et al 1993) and the elimination of this point source drastically
reduced TCDD pollution of the Rhine.

   Of all the possible PCB isomers and congeners, I have chosen PCB 118
(2,3′,4,4′,5-pentachlorobiphenyl) as an example because this is the congener
on which most data are available. In the 1980s the concentration of PCB 118

in the river water in Lobith was around 0.001 μg/l or below. In the sediments of
Lake Ketelmeer a peak was reached between 1960 and 1970 and the level has
steadily decreased ever since (Beurskens et al 1993; Fig. 3B). Concentrations of
about 10 μg/kg sediment were measured in 1990; the mean concentration found
in the suspended matter in 1990 was similar, 8 μg/kg (IRC 1990b). The reduction
of PCB emissions through a world-wide ban is clearly beginning to show results.

Accidental pollution of the Rhine

In addition to the steady contamination of Rhine water through small
uncontrolled effluents (diffuse sources), recalcitrant compounds which pass
through the treatment plants, and the run-offs from roads and fields, accidents
of varying severity contribute to the pollution of the Rhine. Below, an overview
is given of the small accidents recorded in 1990, of the impact of the Chernobyl
reactor accident on the quality of Rhine water and of the Sandoz accident.

Small accidents.   In 1990, 33 small accidents were recorded, of which 13 were
contaminations with different qualities of mineral oil (IRC 1990b). The source
of four oil contaminations could not be ascertained. Five ships were responsible
for the release of about 200 m^3 gasoline; the largest accident happened on
September 13, when 160 m^3 oil were lost. The remaining four accidents
contributed little to the petroleum pollution.

A range of other chemicals were discharged accidentally. In many cases, those
responsible could not be identified. On January 21 there was a discharge of
1,2-dichloroethane, when the highest concentration reached was 4 μg/l. The total
amount discharged and those responsible were not known. Later, 1.4 tonnes aceto-
phenone were released, followed by 0.5 tonnes *o*-methoxyaniline in February,
0.6 tonnes methylpyridine in March, and by 1 tonne methanol, 50 kg chloro-
benzene and 50 kg dichloromethane in April. In May drimarene blue coloured
small parts of the Rhine, and in June, because of the release of 1.3 tonnes of the
herbicide metamitron, Rhine water could not be used as a source of drinking
water for several hours in The Netherlands. In the same month the discharge
of six tonnes of pentachlorophenol into a side-stream killed many fish, though
by the time it reached the Rhine, the chemical had been diluted so much that
it caused no further damage. In August, 250 kg 4-chloro-2-nitroaniline and about
three tonnes of tetrahydrofuran reached the Rhine, followed by three tonnes
of nitrobenzenes in October, and unknown amounts of 3-trifluoromethylalanine
producing a concentration of about 1 μg/l in the Rhine water in November.
The year ended with a contamination of isononanoic acid (highest concentration
25 μg/l) and xylene 14 μg/l in the water on Christmas Eve.

On several occasions treatment plans failed, polluting the Rhine with a
complex mixture of chemicals. Polycyclic aromatic hydrocarbons (PAHs) were
discharged from a ship carrying soil heavily polluted with PAHs. Though they

rarely affect water quality severely, these small accidents are a source of constant, unnecessary pollution. Without the strict year-round monitoring and control of the water composition of the Rhine and the legal measures against polluters, many more such 'small' accidents would probably occur.

The Chernobyl accident.   This accident is a good example of the global impact local events can have. On April 26 1986 an accident in bloc IV of the Chernobyl power plant in the then USSR liberated a large amount of fission products into the atmosphere. The westward air currents brought the radionuclides to the southern Rhine basin on April 30, and by the first days of May to the central and northern parts. The total radioactivity in the Rhine water more than doubled during this period (Fig. 4A). The radionuclides ^{131}I (half-life, $t_{1/2} = 8$ days), ^{134}Cs ($t_{1/2} = 2.06$ years) and ^{137}Cs ($t_{1/2} = 30.17$ years) found in the Rhine water were preferentially adsorbed to particulate matter. When calculated on the basis of dry matter present in the water, the total radioactivity was found to have increased by a factor of more than 10 (IRC 1986). In sediments from Lake Ketelmeer, both ^{134}Cs and ^{137}Cs markedly increased after the Chernobyl accident (Fig. 4B). The first peak of ^{137}Cs, comprising the radioactivity for the late 1950s and early 1960s, is related to the fall-out from atmospheric nuclear weapons testing. ^{134}Cs from this time has obviously already disappeared.

The damage inflicted by the Chernobyl accident on the Rhine's ecosystem was limited by the opening in The Netherlands of all devices which retain the water. This 'flushing' allowed the radioactively polluted water to be rapidly discharged into the North Sea. The passage of Rhine water through sand and gravel during bank filtration and ground-water recharge eliminated most of the suspended solids and as a consequence the excess radioactivity from water subsequently consumed by humans. Though the accumulation of caesium in fish from the Rhine has not been assessed, data from fish living in highly contaminated Rhine affluents (Lippe, at Hamm, Germany) suggested there had been a relatively low contamination (20 Bq ^{137}Cs per kg fish and 10 Bq ^{134}Cs per kg fish, both measured on May 3 and 4 1986; IRC 1986).

The Sandoz accident.   This accident, often regarded as one of the major 'eco-catastrophes', clearly revealed some weak points of surface and ground-water protection, but also clearly showed the river's potential to recover rather quickly after a major poisoning episode.

During a fire on November 1 1986 in a storehouse of Sandoz in Schweizerhalle, near Basle, about 10 000 m^3 of the water which was used to extinguish the fire ran into the Rhine. This water was heavily contaminated because it came into contact with a number of pesticides, in particular disulphoton and thiometon, two phosphoric acid ester insecticides. The water was also contaminated with fungicides, one of which was an organo-mercury compound, herbicides and a number of other chemicals albeit at low concentrations. Between 10 and 30

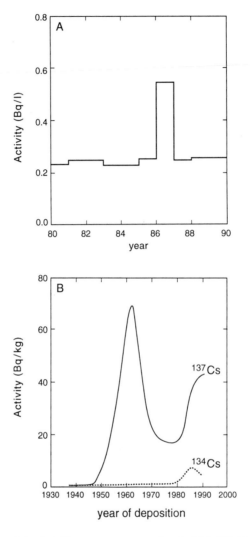

FIG. 4.   Total β-activity in Rhine water measured at Lobith (A) (data from IRC 1988) and ^{137}Cs (——) and ^{134}Cs (----) activities (IRC 1986) measured in the sediments of Lake Ketelmeer (B), according to Beurskens et al (1993).

tonnes, 1–3% of the stored chemicals, ran into the Rhine. The contaminants immediately caused the death of huge numbers of fish and substantial loss of other inhabitants of the water such as mussels, worms, snails, crustaceans and zooplankton.

   Contaminated water ran for about 24 hours into the Rhine, generating a toxic wave of about 40 to 70 km in length. By mixing with waters from affluents,

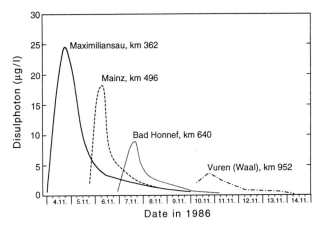

FIG. 5. Change of disulphoton concentrations over time at four measuring stations after the Sandoz accident of November 1 1986, according to IRC (1986). The distance of each station from Lake Constance is given (Rhine km). The accident occurred at Rhine km 159.

the wave spread (Fig. 5) and the chemicals were diluted (Table 2) during the water's 10-day journey to the North Sea 850 km away. Acute negative effects on the fauna could be seen up to around the Loreley, about 400 km from Basle. Further downstream no acute effects could be measured (IRC 1986). The water fauna started to recover relatively quickly after the accident. The open ecosystem allowed the toxic substances to be washed out quite fast, and it allowed an easy invasion by organisms from non-polluted areas which refilled the deserted niches in the formerly poisoned parts of the river. At the moment, it is not possible to draw a final conclusion about the long-term effects of the Sandoz accident on the fauna. It is questionable whether such conclusions ever can be drawn, because extensive and systematic ecological studies of the Rhine started only a few years ago. In addition, the ecotoxicological effects of chronic exposure of the fauna to low amounts of contaminants in the sediments has not yet been extensively investigated. Whether the ground-water along the Rhine has been or will be influenced by the Sandoz accident remains to be seen.

The past, present and future of the Rhine

The River Rhine became polluted early because industrialization and subsequent population growth were centred along its banks during the last century. The Rhine served as source of food, drinking water and energy, was the major water-way for moving goods, and played an important strategic and recreational role. The Rhine was considered an inexhaustible resource, but then fish began to disappear, salmon in particular. The water became increasingly

TABLE 2 Highest concentrations of contaminants measured in the Rhine after the Sandoz accident on November 1 1986

Place	Distance from Lake Constance[a] (Rhine km)	Date (1986)	Time of day	Disulphoton (µg/l)	Thiometon (µg/l)	Etrimphos (µg/l)	Propetamphos (µg/l)	Oxadixyl (µg/l)	Ethylparathion (µg/l)	Mercury (µg/l)
Märkt	173.00	01.11	15.15	600	500	50	100	80		12
Wyhl	244.35	02.11	16.45	107	23	10	6	37	1	2.6
Gambsheim	310.00	04.11	00.00–03.00	73	15	4	2	32	0.5	0.6
Maximiliansau	362.00	04.11	12.00–24.00	24.6	10.6	3.1	1.1	11.5	0.4	
Ludwigshafen	428.00	05.11	10.33	30.3	14.4	3.1	1.0	12.3	0.4	0.4
Mainz	498.00	06.11	04.30–08.30	18.3	8.3	2.6	3.4		0.4	
Koblenz	590.00	07.11	02.00–04.00							0.2
Neuwied	609.00	07.11	09.00–10.00	11.8	3.9	1.3	0.6	6.9	0.3	
Bad Honnef	640.00	07.11	14.00–18.00	8.9	3.5	1.1	1.0		0.1	
Düsseldorf	734.00	08.11	08.15	5.7	2.2	0.7	0.6		<0.1	
Lobith	862.30	09.11	09.00	5.3	2.0					0.22
Vuren (Waal)	952.00	10.11	19.00	3.3	1.3					0.08

[a]The accident occurred 159 km from Lake Constance.
Data from the International Rhine Commission (IRC) 1986.

contaminated with pathogens which made the river less attractive for recreation and forced authorities to take specific measures for drinking water production. Finally, many of the new synthetic chemicals, their intermediates and the compounds from which they were synthesized severely spoiled the quality of the water. In the late 1950s and 1960s the River Rhine was Western Europe's open sewer.

A number of commissions had and still have to control the pollution of the Rhine; among them the most important is probably the International Commission for the Protection of the Rhine Against Pollution, founded in 1950. This commission concentrates particularly on determining the origin and importance of different types of pollution and makes proposals for inter-governmental agreements for the reduction of pollution. Figures 2 and 3 clearly show the substantial positive effects this commission has had. According to the conceptual model of pollution of Meybeck et al (1989), the Rhine is clearly in the recovery phase (IV, Fig. 6). It is evident that more efforts are needed to reduce

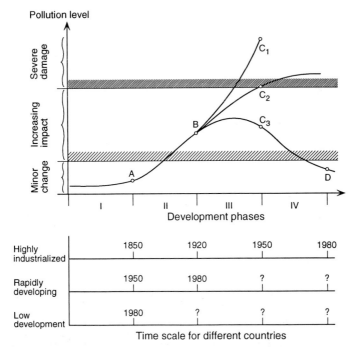

FIG. 6.   A conceptual model of pollution and its control. Sewage pollution in Europe is taken as an example (redrawn from Meybeck et al 1989). The phases labelled A–D are: A, pollution increasing linearly with increasing population; A–B, pollution increasing exponentially with industrialization; B–C_1, no pollution control enacted; B–C_2, some pollution controls enacted; B–C_3, effective controls consistently employed; C_3–D, remedy of the pollution to an environmentally tolerable level, due to effective control of pollution at source and/or advanced treatment procedures.

pollution further if the Rhine is to maintain its present roles and regain those things it has lost. The 'Action Programme Rhine', which has been set up by the governments of France, Germany, Luxemburg, The Netherlands, Switzerland and the European Commission, has the following goals, to be achieved by the year 2000:

Higher animals, such as salmon, which were indigenous should return to the Rhine.
Rhine water should continue to serve as source of drinking water in the future.
Pollution of the sediments should be reduced.
The Rhine should again function as balanced ecosystem, and should not be allowed to further endanger the ecosystem of the North Sea, neither continuously nor accidentally.

These four goals aim for more than merely pollution reduction—strict ecological components have been included. The development of an ecological concept will become a major task in the future because the geomorphology of the Rhine basin has changed drastically over the last 150 years, as shown for the Basle region in Fig. 7; the built-up areas have grown considerably, the river has been straightened and formerly open land is now inhabited or used for agriculture. A successful development and application of ecological practices

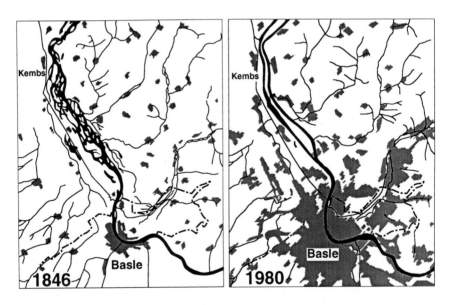

FIG. 7.   Development of the Rhine and the city of Basle and its environs between 1846 (*left*) and 1980 (*right*). The shaded zones are built-up areas. (Figure courtesy of R. Koblet, unpublished.)

is needed; without this, sustainability will remain an empty word, and the Rhine will not become the 'spring of life' for Western Europe again.

The future elsewhere

The more advanced countries, mainly those in Europe and North America, have experienced dramatic pollution events in their recent histories. They have undertaken serious actions to mitigate and curb water pollution, as discussed above for the Rhine. In countries with a low level of development, point-source pollution is only a minor problem because sewerage networks and heavy industries are lacking. Water pollution in cities may be severe, however. These societies are in phase II, between A and B on the curve in Fig. 6.

Pollution threatens rapidly growing and industrializing countries such as Brazil, China, India, Indonesia, Mexico and Nigeria, where demands upon water resources are increasing exponentially. Only 10 out of 60 countries in this category have established effective laws, regulations and enforcement policies to cope with the growing pollution problem (Helmer 1987). Water pollution which appeared over a period of more than 100 years in Europe occurs in such rapidly developing countries over a period of much less than one generation (Fig. 6). Achieving control over water pollution in such countries is necessary not only for long-term ecological and economical sustainability but also for mere immediate survival.

Acknowledgements

My special thanks go to Dr Rudolf Koblet, who helped me to collect some of the data and allowed me to use one of his unpublished figures (Fig. 6). The figures were drawn by Heidi Bolliger and the manuscript was typed by Gerda Thieme.

References

Ardern E, Lockett WT 1914a Experiments on the oxidation of sewage without the aid of filters. I. J Soc Chem Ind 33:523–539
Ardern E, Lockett WT 1914b Experiments on the oxidation of sewage without the aid of filters. II. J Soc Chem Ind 33:1122–1130
Beurskens JEM, Winkels HJ, Mol GAJ, Barreveld HL, van Munster G 1993 Geochronology of priority pollutants in sedimentation area of the Rhine River. Environ Toxicol Chem, in press
Helmer R 1987 Socio-economic development levels and adequate regulatory policy for water quality management. Water Sci Technol 73:257–272
ICSU 1992 The Dublin statement on water and sustainable development. Sci Int 47:9
IRC 1986 Annual report. International Commission for the Protection of the Rhine against Pollution, Koblenz

IRC 1988 Annual report. International Commission for the Protection of the Rhine against Pollution, Koblenz

IRC 1990a Statistical tables for 1990. International Commission for the Protection of the Rhine against Pollution, Koblenz

IRC 1990b Annual report. International Commission for the Protection of the Rhine against Pollution, Koblenz

Meybeck M, Chapman D, Helmer R 1989 Global fresh water quality: a first assessment. World Health Organization and the United Nations Environment Programme. Blackwell Scientific Publications, Oxford

Nriagu JO (ed) 1980 Cadmium in the environment. I. Ecological cycling. Wiley, New York

Nriagu JO (ed) 1981 Cadmium in the environment. II. Health effects. Wiley, New York

Rappe C 1980 Chloroaromatic compounds containing oxygen. In: Hutzinger O (ed) Handbook of environmental chemistry. Springer-Verlag, Berlin, vol 3A: 157–179

Safe S (ed) 1987 Polychlorinated biphenyls (PCBs): mammalian and environmental toxicology. Springer-Verlag, Berlin (Environ Toxin Ser 1)

WHO 1977 Lead. World Health Organization, Geneva

WCED 1987 Our common future. Oxford University Press, Oxford

WRI 1986 World resources 1986: an assessment of the resource base that supports the global economy. Basic Books, New York

DISCUSSION

Hoffmann: What is the extent of primary, secondary and tertiary waste water treatment along the River Rhine, and is the Rhine used as the discharge point for waste water?

Zehnder: There is an action programme not only for the River Rhine but also for the North Sea. The aim is to reduce the pollution by phosphates and nitrogen to 50% of the 1989 levels by 1995, and to 75% by the year 2000. Of all the countries along the Rhine, Switzerland and Germany were probably the leaders in waste water treatment. When I came to The Netherlands 10 years ago I was surprised how few waste water treatment plants there were; waste water was largely discharged into the River Rhine or directly into the North Sea. The situation is now changing. Switzerland has probably done most. In Switzerland there is now almost no untreated water discharged. There is secondary treatment of most waste water and an increasing volume is subjected to tertiary treatment. In the other countries, large amounts of waste water receive primary treatment but still need secondary treatment. If the goals of the North Sea Action Plan are to be met by 1995, there must be at least secondary treatment of all waste water. For the goals of 2000 to be achieved, tertiary treatment of almost all waste water is essential.

Klein: There are forms of pollution to consider other than the obvious chemical ones. Nuclear power stations produce thermal pollution, which may affect oxygen concentrations. There is also the problem of salt pollution from the salt mines in France, which is not covered as a chemical pollutant in the usual way. Also, we should consider the traffic, whether all the ships moving up and down the river have sewage treatment systems, and the problems of spillages and so forth.

Zehnder: The temperature of the River Rhine has increased by about 2 °C over the last 15 years. This might be the result of the enhanced greenhouse effect, but thermal pollution could also contribute. A more detailed analysis of the sources needs to be done.

The French government has introduced measures to reduce the release of salt into the River Rhine to a level low enough to enable the water to be classified as fresh water. Raised salinity would render Rhine water brackish and consequently of limited use for drinking water production. The second point of the Rhine Action Programme states that the quality of Rhine water should be such that it is suitable for production of drinking water. This point in the Rhine Action Programme actually defines the limit of salinity.

The river traffic isn't a tremendous problem as long as there are no accidents. Many of the oil accidents involve leakage and the problems caused are less severe than those resulting from failure of sewage treatment plants or from accidents in chemical industries. Regulatory measures can reduce water pollution. The River Rhine is quite heavily controlled by a 'police force' chasing polluters. Without such strict control there would be more so-called accidents. The quality of the water is constantly monitored, and when accidents happen immediate actions are taken.

Mansfield: When you refer to a balanced ecosystem, do you mean a community of plants, animals and microorganisms that can tolerate a certain level of man's activity?

Zehnder: A balanced ecosystem is difficult to define. By this I mean mainly animals and plants, because microorganisms are more adaptable and their diversity is probably higher in polluted areas than in non-polluted areas. A balanced ecosystem could be defined as an ecosystem composed of populations of plants and animals which would normally live along the river and in the water if it were not polluted. In short, the ecosystem should not be negatively affected by pollution. This is what the Rhine Action Programme aims for.

Mansfield: Would you re-establish some of the organisms known to have been there in the past?

Zehnder: Yes; higher animals, such as salmon, have to come back. The salmon is probably more of an indicator organism than anything else. There are problems other than water quality in returning salmon to the river, such as the barriers and power plants along the river which prevent their migration. The return of the salmon is the translation of a scientific goal into one which can be understood by the public.

Mansfield: It's a symbol of cleanliness, essentially.

Zehnder: Yes. It's thought that if the salmon can survive, so can all the other animals.

Kleinjans: Salmon act as an indicator for pollution by heavy metals, predominantly. The increasing loads of phosphates and nitrates in combination with the increase in temperature will affect the ecosystem, but the salmon could still live in the water. How isolated within the framework of environmental policy is this goal of the return of the salmon? Is there a hope that the other higher organisms will follow once the salmon has returned?

Zehnder: There is no guarantee. The salmon is merely symbolic.

Rabbinge: The salmon is an indicator *and* a symbol.

de Haan: How can the Rhine Action Programme improve water quality as far as heavy metals are concerned? What is the contribution of road traffic to lead and cadmium in the river water? In The Netherlands we have every year to dispose of millions of cubic metres of sludge produced when harbours are dredged, and we now can't find safe storage facilities because this soil is contaminated with organic compounds. This situation certainly will be improved by the Rhine Action Programme, but the problem of heavy metals, specifically lead, cadmium and arsenic, remains. The traffic in the river basin area is the main contributor to this problem. Is this pollution tackled in the programme?

Zehnder: This is covered by the third point, the reduction of pollution of the sediment. We could, for example, drastically reduce the input of lead into the River Rhine by removing lead from petrol.

de Haan: That measure is not included in the River Rhine Action Programme. It is something each country has to do on its own.

Zehnder: That's correct. The plan consists of general goals set by the IRC. The IRC, which was founded in 1950, is composed of representatives from Switzerland, Germany, France, Luxemburg, The Netherlands and the EC. As a result of its work, pollution has decreased since 1960. The long-term goals of the commission may not have an immediate effect, but might influence future political decisions.

Avnimelech: You mentioned low and high per capita availability of water, defining high availability as above $10\,000\,m^3$/person/year. In Israel, availability is $300\,m^3$/person/year, so we have to manage our water efficiently. In the early 1970s there were indications that in the Sea of Galilee, which is downstream of the River Jordan, the process of eutrophication was beginning. A few scientists approached the media to raise public awareness. Public opinion forced the Government to act, and they established a river basin authority, a coordinated authority in which industry, the municipalities, the Government and agriculture were represented. To cut a long story short, all development had to be controlled to prevent pollution of the water. The quality of the water in the River Jordan and in the lake has been improved tremendously. An economic analysis showed that the benefits of this effort covered the expenses. The conclusion from this is that public opinion is essential for such a project, as is a coordinated and comprehensive approach. You cannot deal with the issue of water by looking at only one topic in isolation. It is possible to succeed, and if you succeed it can be economically profitable.

Lake: The Rhine Action Programme has four action points, and only one of those relates explicitly to human health. Would addressing that one only, making the Rhine water potable at all stages, entrain the other three? If Rhine water is potable, safe for us to drink, would the other three criteria be met? To me, one of the attractions of this meeting was that it stresses that human health comes first, in contrast to pressure from 'green' groups to protect natural ecosystems regardless of the implications for humans.

Zehnder: This is a tricky question. Technically, we are able to make drinking water out of almost anything which still contains water. The programme does not define whether water should serve as drinking water directly or after bank and dune filtration.

Bradley: In a country like Togo or Indonesia, the data, certainly those from East Africa, suggest that less than $2\,m^3$/person/year of water is required for individual domestic use, meeting the needs for life and health. It is worth bearing in mind that quite modest goals in relation to developing countries can have major health benefits.

Zehnder: We have to bear in mind that these countries also wish to develop industrially, which requires a higher water consumption, and their agriculture may also need more water. $300\,m^3$/person/year, the availability in Israel, is considered to be low.

Edwards: Is there any place for artificial oxidation or aeration of rivers?

Zehnder: Rivers aerate themselves quite efficiently through turbulence. To my knowledge, the River Rhine, even at the sediment level, has never become anaerobic. The total concentration of oxygen was lower in the 1950s when there was almost no waste water treatment and the river actually functioned like a plug flow waste water treatment plant. The average oxygen concentration in the Rhine has increased in the last 10 years by about 2–3 mg/l.

Assessing the greenhouse effect in agriculture

R. Rabbinge*†, H. C. van Latesteijn* and J. Goudriaan†

*Netherlands Scientific Council for Government Policy (WRR), PO Box 20004, 2500 EA The Hague and †Department of Theoretical Production Ecology, Wageningen Agricultural University, PO Box 430, 6700 AK Wageningen, The Netherlands

Abstract. Evidence that concentrations of CO_2 and trace gases in the atmosphere have increased is irrefutable. Whether or not these increased concentrations will lead to climate changes is still open to debate. Direct effects of increased CO_2 concentrations on physiological processes and individual plants have been demonstrated and the consequences for crop growth and production under various circumstances are evaluated with simulation models. The consequences of CO_2 enrichment are considerable under optimal growing conditions. However, the majority of crops are grown under sub-optimal conditions where the effects of changes in CO_2 are often less. The same holds for the possible indirect effects of environmental changes such as temperature rise. Studies on individual plants under optimal conditions are therefore not sufficient for evaluating the effects at a farm, regional, national or supra-national level. Simulation studies help to bridge the gap between the various aggregation levels and provide a basis for various studies of policy options at various aggregation levels.

1993 Environmental change and human health. Wiley, Chichester (Ciba Foundation Symposium 175) p 62–79

The effects on plant growth of changes in ambient CO_2 have been investigated under experimental conditions. However, the effects on agricultural production cannot be assessed on the basis of these experimental results. Agricultural production is susceptible to a large number of environmental changes. Soil, air and water pollution as well as soil degradation and erosion may all cause a decline. The threat of global climate change has added another major concern to this list. Most research efforts, however, are directed towards understanding physiological effects at the subcellular level or at the level of individual plants. In these studies it is assumed that plants grow under optimal conditions. Most productive plants grow under circumstances where some of the conditions for optimal growth are not satisfied; for example, there may be shortage of water or nutrients or there may be pest infestation, and growth is restrained by one of these conditions. In such situations the effects of climate change may be considerably smaller than those predicted on the basis of research results under experimental conditions.

Changes in CO_2 are only one aspect of climate change. Changes in temperature, radiation and precipitation may also occur, complicating the assessment of the net effects. The direct effects of these changes on crop growth-defining factors and indirect effects exerted through growth-limiting and growth-reducing factors are investigated at the level of individual plants and at the crop level. The possible effects of climate change on agriculture as a socio-economic sector operate at farm, regional, national and supra-national levels, where the consequences of other changes may be much more important. To assess the net effects of climate change an integrative methodology is necessary. This can be provided by a multiple goal linear programming technique designed to integrate the various changes and to evaluate their relative effects. The results of this assessment can be used to help in the development of policy options.

Environmental conditions and environmental changes

For many centuries there have been slow but persistent changes in atmospheric CO_2 concentration and temperature. Changes, as such, are therefore normal and should not lead to any problems. However, the change of ambient CO_2 concentration which occurred during the last century was bigger than that of the previous 10 000 years. This rapid change has also been found for various other trace gases, such as methane (CH_4), nitrous oxide (NO_x) and ozone (O_3). Since pre-industrial times the concentrations of these trace gases have increased by between 15% and 200%. Chlorofluorocarbons (CFCs) are an entirely new, artificial addition.

It is likely that this trend will continue at a similar or perhaps even faster pace in the future. Since the end of World War II annual emissions of CO_2 have increased by more than 300%. The rates of increase have been especially high in the People's Republic of China and developing countries in south Asia, Africa and Latin America. By 1990 these countries were responsible for more than 25% of the global emission, whereas in 1950 they were responsible for about 7%. The relative contribution of these areas will increase considerably during the coming decades as a result of population growth, a tremendous increase in combustion of fossil fuel, and deforestation. If all fossil carbon available for combustion is burned, the ambient CO_2 concentration will increase up to a maximum of about 2000 p.p.m. This will take about 200 years. After that, the CO_2 concentration will decrease again through geological and biological processes. Although the consequences on a human time-scale might be devastating, on a geological time scale this is merely a ripple on the water.

The increase in atmospheric CO_2 may intensify the greenhouse effect, which may lead to a rise in overall mean temperature world-wide. This prediction is not based on extrapolation of current trends, but on energy balance climate models.

The outcomes are uncertain and the size of the projected temperature increase differs between models, but virtually all the models predict an increase. It seems that temperature will increase, but the magnitude is uncertain. Even more uncertain are predictions about rainfall, cloudiness and radiation intensity in various places.

Agricultural production

Agriculture may be defined as the human activity that produces primary products with the sun as the major source of energy. The role of agriculture has changed considerably since its beginnings. Over the last century particularly there has been a significant increase in productivity per unit of area and per person. Since World War II there has been a sharp rise in productivity growth in all agricultural areas as a result of better manipulation of agricultural factors. On average, the productivity growth used to be 3 to 4 kg/ha/year and changed to more than 50 kg/ha/year in a very short period.

This change has been found everywhere, independent of the socio-economic system or climate zone. The combination of knowledge from various disciplines and the new methods, especially the application of external inputs such as plant nutrients and pesticides, help to increase productivity. In parts of the industrialized world productivity has risen from 1500 kg wheat (*Triticum* spp. L.) per ha early this century to 8000 kg/ha at present. In about 80 years labour productivity improved from 200 hours/tonne to 2 hours/tonne.

Yields may be considered at different levels: potential, attainable and actual (Rabbinge 1986). The *potential* yield is determined solely by yield-defining factors and the physiological, phenological, geometrical and optical characteristics of the crop. The physiological characteristics determine the way light energy is used to produce the sugars which constitute the basis for all structural materials in the crop. The geometrical and optical characteristics determine what fraction of the incoming radiation is intercepted and absorbed. The phenological characteristics determine, according to temperature, the rate at which various development stages are passed.

The *attainable* yield is lower because crop growth-limiting factors such as nutrient or water shortage occur. During a part of the growing season water or nutrients may be restricted, resulting in a decrease in yield. More than 90% of agriculture takes place under such circumstances. The *actual* yield is even lower as a result of crop growth-reducing factors such as pests, diseases, weeds or air pollution.

The yield-defining factors determine the potential for primary production. They can be affected or changed little by management decisions, because environmental conditions such as incoming radiation, temperature and CO_2 concentration can not be manipulated by the individual farmer. The yield-limiting factors (water and nutrients) as well as the yield-reducing factors (pests)

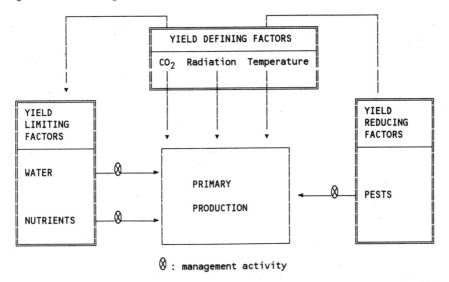

FIG. 1.   The influence of climatic variables on agricultural production. Stages at which management decisions can have an effect are indicated (⊗).

can be influenced by management decisions. The interactions between these factors are schematically represented in Fig. 1.

An increase in atmospheric CO_2 will have a primary physiological effect on production by increasing photosynthetic activity (Lemon 1983, Goudriaan 1990). Although the initial positive response to an increased CO_2 concentration has been demonstrated convincingly, there are uncertainties about long-term effects.

In addition to this primary effect, changes in temperature, radiation and precipitation may have secondary effects on plants. An increase in temperature would speed up development, thus shortening the growth period, and would therefore have a negative effect on yields for determinate crops (those which have a clear period of productive growth, such as wheat) (Nonhebel 1990), but would probably increase yields of indeterminate species (those which grow productively over nearly the whole growing season, such as grass) (Squire & Unsworth 1988). Changes in radiation can complicate this effect. Increased cloud cover might have a negative effect on production during the growth period. Uncoupling of present temperature–photoperiod complexes would probably lead to changes in the geographical distribution of perennial species (Habjorg 1990).

A change in precipitation (temporal or geographical) can have positive or negative effects on local production potentials, depending on the shifts in water availability (Rabbinge 1986). This can of course be compensated by management responses, such as an increase in irrigation or drainage activities. A temperature rise combined with changes in water availability can affect availability of nutrients because of changes in mineralization in the soil which can induce deeper drainage and possible leaching of nitrates. Finally, as a result of changes in temperature, pests can shift. For example, the cornborer, *Ostrinia nubilalis*, will tend to spread more to the northern regions, introducing new problems in maize production that require changes in pest management. The net effect of all these changes cannot be predicted with current knowledge (Parry 1990).

Direct effects of climate change on photosynthesis

There is a great deal of experimental evidence available on the effect of ambient CO_2 concentration on photosynthesis and the subsequent accumulation of dry matter in crops (Lemon 1983, Strain & Cure 1985, Kimball 1983, Cure & Acock 1986). Light is indispensable for the process of photosynthetically driven CO_2 uptake by green plants. In the natural environment, light and CO_2 are normally sub-optimal; consequently, photosynthesis is stimulated by an increase of ambient CO_2 (Fig. 2), not only under high light intensity, but also under low light conditions.

There are clear differences between two major classes of plants, C_3 and C_4 plants, which differ in biochemical and anatomical aspects in the way they take up and utilize CO_2 from the ambient air. In C_3 plants the enzyme that binds CO_2 can also be inhibited by O_2 (Farquhar & von Caemmerer 1983). Enzyme bound to O_2 must be recovered. This recovery costs energy and releases CO_2, observable as photorespiration. Because CO_2 and O_2 compete for the same site on the enzyme, photorespiration is suppressed by higher CO_2; higher CO_2 concentrations will lead to a higher proportion of CO_2 binding the enzyme and thus to a higher rate of photosynthesis. In C_4 plants, mostly tall tropical grasses such as millet (*Panicum miliaceum* L.), maize (*Zea mays* L.), sorghum (*Sorghum bicolor* [L.] Moench) and sugar-cane (*Saccharum officinarum* L.), the enzyme which binds CO_2 does not bind O_2. Higher levels of ambient CO_2 will therefore have no effect on the process of photosynthesis through this route. However, photosynthetic activity may be influenced by partial stomatal closure, a typical secondary effect of an increase in ambient CO_2. Stomatal pores in the epidermis of leaves are necessary for the uptake of CO_2 from the ambient air, but at the same time water vapour escapes (transpiration). The degree of opening can be considered as a compromise in the balance between limitation of water loss and admission of CO_2. The generally observed closure of stomata when CO_2 increases is an expression of this compromise (Wong 1979). The much higher affinity of the CO_2-binding enzyme for CO_2 in C_4 plants permits them to maintain a more favourable ratio between net CO_2 uptake and evaporation of water.

If ambient CO_2 is raised, net CO_2 assimilation may be increased and water loss may be reduced, depending on how the stomata react. Figure 2 shows the CO_2 supply–demand function of assimilation. The figure shows how the assimilation rate is affected by the CO_2 concentration in the stomatal cavities

(a) C_3 species response

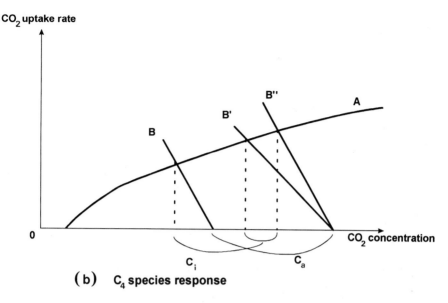

(b) C_4 species response

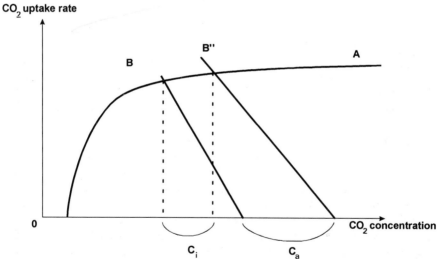

FIG. 2.   The response (A) in (a) C_3 plants, which show a strong assimilation response to CO_2, and (b) C_4 plants, which show an assimilation response only at a low CO_2 concentration, of leaf photosynthesis to the concentration of CO_2 inside the stomatal cavity (C_i). The rate of photosynthesis affects C_i through stomatal supply limitation (B). Lines B' and B" indicate the effects of a doubling of [CO_2] with no stomatal response (B') and with partial stomatal closure (B") resulting from the increase in CO_2. Adapted from Goudriaan & Unsworth (1990).

(line A). This intercellular concentration decreases linearly from the ambient CO_2 concentration when the assimilation rate increases (line B), because of the resistance offered by the stomata. The point of intersection of the two lines (line B with line A) gives the realized rate of assimilation and intercellular CO_2 concentration. Raising external CO_2 will increase assimilation to a maximum (line B') if the stomata do not respond at all, but the stomata usually close somewhat (line B''). In an extreme situation where the intercellular CO_2 concentration is kept constant, when ambient CO_2 increases the stomatal closure will be maximal; as a result there will be no effect on assimilation. However, the most common response is that the ratio of intercellular (C_i) to external CO_2 (C_a) concentrations is stabilized, thereby balancing CO_2 assimilation and water loss. This constant $C_i : C_a$ ratio means that the stomatal aperture is reduced very little if assimilation responds strongly to C_i (typical in C_3 plants). In C_4 plants photosynthetic CO_2 uptake is usually almost saturated, and stomatal closure is maximal (Fig. 2b). The combined effect can be summarized as follows: typically with a doubling of C_a, in C_3 plants transpiration will be reduced by 10–20% and assimilation stimulated by 40%, and in C_4 plants only transpiration will be reduced, by up to about 25%. In both types this leads to a considerable increase in efficiency of water use.

Effects of an increase in ambient CO_2 concentration on crop production

Net effects under optimal conditions

During the growing season, several internal mechanisms operate in the plant, modifying the initially observed effects of CO_2. In plants adapted to high CO_2, photosynthesis per unit leaf area is often smaller (Wong 1979, Mortensen 1983) than in non-adapted plants (when measured under equal circumstances), but in soy bean (*Glycine max* [L.] Merr.) increased photosynthetic capacity has been observed (Valle et al 1985, Campbell et al 1988). When grown and at a higher CO_2 concentration, leaves generally have a higher rate of photosynthesis. This is particularly true for N_2-fixing plants, which have nodules in their rooting system. The growth response of these plants to CO_2 tends to be particularly strong.

Starch accumulation (Ehret & Jolliffe 1985) tends to cause some increase of leaf weight per unit leaf area. This rather passive response will not increase light interception, but the more active response of formation of more leaf area, by producing larger leaves or more tillers, will further enhance the effect of increased CO_2 at the level of a whole crop.

As shown in a review by Kimball (1983), responses to CO_2 at the single leaf level are carried over to crop yield; the response to a doubling of CO_2 concentration is a mean increase in dry matter of 40% for C_3 crops and 15% for C_4 crops.

Net effects with water shortage

The effect of raised CO_2 under good growing conditions is maintained fully under conditions of water shortage. Gifford (1979) and Sionit et al (1980) have shown that growth, yield and efficiency of water use of wheat under water shortage can be considerably improved by raised ambient CO_2. Kimball et al (1986) reported that a doubling of CO_2 concentration led to relative increases of cotton (*Gossypium hirsutum* L.) production ranging between 50% and 70% with both optimal and limiting levels of water supply. Similarly, Goudriaan & Bijlsma (1987) showed that efficiency of water use in Faba bean (*Vicia faba* L.) was improved by about 50% under doubled CO_2, with both normal and limited water supply. For a C_4 grass, however, Gifford & Morison (1985) found that growth was stimulated by CO_2 only under severe water stress.

In conditions of high salinity plants grow continually under osmotic stress. Schwarz & Gale (1984) found that the CO_2-induced increase of dry matter in saline conditions was equally strong as or even more pronounced than in non-saline situations, and that C_4 halophytes (plants that are salt tolerant) responded as strongly as C_3 halophytes.

A major reason for the strong positive effect of CO_2 in plants subjected to water stress is the common physical pathway of water vapour and CO_2 through the stomatal pores. The effect might partly be explained by better availability of assimilates with which to make osmotic adjustments and thereby maintain turgor. These results are especially important for arid regions, where brackish or even saline conditions are common. Even when a limited stock of water is available, the plants may have adequate water for most of the growing period. The water shortage occurs only when the soil has lost about 75% of its initial water content. An increased efficiency of water use will result in a greater biomass in proportion to the amount of water used, but the amount of water left in the soil will be unchanged. However, if the plants are also limited by nutrient shortage or by the duration of the development period, the increase in biomass will be too small to compensate for the decreased transpiration rate and the amount of water left in the soil after the growing period may increase.

Net effects with nutrient shortage

Nutrient shortage tends to limit crop growth more than water shortage, and without leaving much room for stimulation by CO_2. Increased accumulation of starch in leaves grown under high CO_2 gives a general increase in dry matter. Shortage of phosphorus was found to limit growth almost independently of CO_2 in a pot experiment (Goudriaan & de Ruiter 1983). Also, the growth-limiting effect of potassium shortage is not alleviated by higher CO_2 (J. Goudriaan, unpublished data 1985). In open soil and perennial species a positive effect of increased CO_2 concentration might occur through more intense rooting and increased soil weathering (Rosenberg 1981).

Norby et al (1986) could not find a disappearance of the positive effects of a CO_2 increase, even at very low nutrient levels. However, the effect of CO_2 enrichment under such conditions is much less, because nutrient shortage limits growth. Nitrogen appears to be in an intermediate position; increased CO_2 has a small positive effect even under rather severe nitrogen shortage (Goudriaan & de Ruiter 1983). In addition to increased starch accumulation, which generally lowers leaf nutrient contents, there is also a decrease in the carboxylation enzyme rubisco (ribulose-bisphosphate carboxylase), which contains up to 50% of leaf nitrogen. In accordance with this photosynthetic role of nitrogen, the nitrogen content of leaf tissue is lower in plants grown under high CO_2 than under low CO_2. In seed tissue, however, the C:N ratio is more stable (Kimball et al 1986).

An intriguing question is how canopy transpiration is affected by CO_2 when plant growth is nutrient-limited. Lenssen (1986) showed that the reduction in transpiration caused by an increase in CO_2 was the same under severe phosphorus limitation. This result is relevant with regard to water yields from watersheds with naturally growing vegetation. These natural ecosystems are quite often strongly nutrient-limited. Growth is not improved under conditions of increased CO_2 because of nutrient limitations, but an increase in the water yields may be expected.

Net effects of changes in temperature and the growing season

As indicated above, with a temperature rise the rate of development of crops may increase and the length of the growing season will be shortened, especially for determinate crop species. Unless a temperature rise is very high, the decrease in yield it causes is more than compensated by the effect of CO_2 enrichment. The effects of a temperature rise on pests and diseases are very uncertain because primary pests may increase in number but so might their natural enemies.

The influence of other climatic factors

Other climatic factors such as precipitation and radiation are no less important than temperature (Rosenberg et al 1990). Unfortunately, there is even more uncertainty about the way they will change than there is about the likely change in temperature. Crop growth models show that potential productivity is closely related to incoming radiation during the growing season, but the general circulation models (GCMs) that are used for climate forecasting are not yet able to produce reliable predictions on this point. The possible impact of radiation changes is therefore usually ignored, in spite of their potential importance.

The current prediction is that mean global precipitation will increase by 7–15% (Wilson & Mitchell 1987). Some regional specificity can be obtained from maps such as those produced by Schlesinger & Mitchell (1985). A model for wheat and rice that included both climatic changes (temperature and precipitation) and the physiological effects of doubled CO_2 predicted that potential crop yields would increase by 10–50% at some sites in Europe and Asia (van Diepen et al 1987). The mean GCM scenarios were superimposed on current weather, retaining current variability. The resulting variability in yield was less than found presently, so more stable wheat yields can be expected.

The effects of climate change at the level of individual farms, regions and countries, and implications for policy-making

As a result of accelerating changes in technology and its dissemination, changing relations within the global market for agricultural products and adaptations in policy, the agricultural sector will show major structural alterations in the near future. The implications of climatic changes for management and policy cannot easily be singled out. A clear understanding of the overall developments in agriculture is indispensable for the formulation of appropriate policy reactions to climate changes. This is illustrated by the increase in the world-wide average grain yield per unit area that occurred between 1959 and 1986. In this period the yields increased from 1400 kg/ha to about 2600 kg/ha (Food and Agriculture Organization of the United Nations [FAO] 1987). This near doubling was the result of a continued annual relative growth rate of about 2.3% per year. Over the same period, atmospheric CO_2 rose from 315 to about 345 p.p.m., at an average rate of about 0.34% per year. When yields over this period are plotted against atmospheric CO_2 concentrations, an almost perfect correlation is evident (Fig. 3). It would be very unwise indeed to draw any conclusions from this graph about cause and effect, but it does raise some interesting considerations.

Figure 3 shows that the relative growth rate of cereal production (on an area basis) has been seven times as large as the increase in atmospheric CO_2. We know from experiments that under optimal conditions for growth $1/14$ at most of this increase in crop production can be reasonably ascribed to the increase in atmospheric CO_2. This relatively small impact of the rise in CO_2 strengthens the point of view of those who consider the climatic change issue to be minor in comparison with other opportunities and threats of human origin.

The yields from individual farms have increased considerably over the last decades, mainly because of improved farm management and the proper use of external inputs. There is still ample room for further rises in actual yields; in many cases, the actual yield is only 10–20% of the potential yield. Whether or not these potential yields will be achieved is dependent to a great extent on policy decisions. It is for that reason that the assessment of the consequences of climate change for the agricultural sector should include the options for future developments of agriculture at a regional, national or supra-national level.

One way to provide this sort of information is to assess the quantitative relations between changes in climatic conditions and a number of self-contained

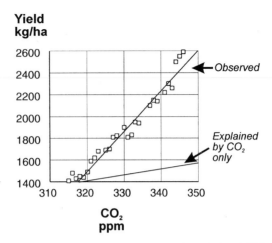

FIG. 3. Average world-wide grain yields (Observed) and the yield increase calculated to result from increased ambient CO_2 between 1959 and 1986 versus atmospheric CO_2 concentration measured at Mauna Loa, Hawaii. Adapted from Goudriaan & Unsworth (1990).

technical developments in agriculture. These technical relations limit the possibilities for future developments. Next to that a set of objectives can be identified which play an important role in the policy debate on the future of agriculture. These objectives are related to agricultural production itself, but also to other socio-economic aims, criteria with respect to environmental protection and requirements for nature conservation. The consequences of interactions and preferences among the different objectives for rural areas can then be examined, given the technical possibilities and the expectations for climatic conditions. The large number of uncertainties make it impossible to construct a model that can predict changes in agriculture resulting from changes in climatic conditions. However, it is possible to construct a model that provides consistent information based on 'what if' questions. This methodology has been developed by the Netherlands Scientific Council for Government Policy and has been used to explore options for future land use in the European Community (Netherlands Scientific Council for Government Policy 1992, Rabbinge & van Latesteijn 1992; van Latesteijn 1993). In this way, the relative importance of the effects of climate change on agriculture in comparison with changes that will result from policy changes can be assessed. For example, it might very well be that changes in agriculture related to environmental protection and nature conservation issues turn out to be much more profound than changes resulting from climate change.

  Policy decisions intended to oppose the possible adverse effects of climate change must always be regarded in this broader context. It might be that compared with other issues climate change has only a minor impact on policies.

Although there will always be a discrepancy between the policies suggested by results of rational research and the policy-making process in real life, assessment of the relative importance of climate change is a prerequisite for sound policy planning.

Coupling crop growth simulation models with land use models and providing this system with information on climate changes is one way in which to construct a methodological framework that can be used to assess the relative effects of climate change on agriculture. In a quantification of the possible developments and their interactions, land use must be the central theme; through changes in land use all other changes can be linked to each other. On the basis of soil characteristics, climatic conditions and the properties of the crop, regional yield potentials for indicator crops can be calculated using a dynamic crop simulation model and a geographical information system. Next, the influence of climate changes and policy preferences on the regional allocation of production can be assessed. In this way, the relative contribution of climate change to the available options for agriculture can be explored. This procedure is outlined in Fig. 4.

In general, the effects of climate change may be of minor importance. Nevertheless, this methodology may help us to gain better quantitative insight into the present options for land use and the changes resulting from climate change. In this way, the combination of biotechnical studies, which provide the basic information needed on all plant and crop levels, and socio-economic studies on objectives and constraints may generate land-use scenarios for long-term policies.

Conclusions

The effects of change in ambient CO_2 on growth and production of crops growing under optimal conditions are considerable. Production may increase, water use efficiency improve and yield variability decrease. However, most crops

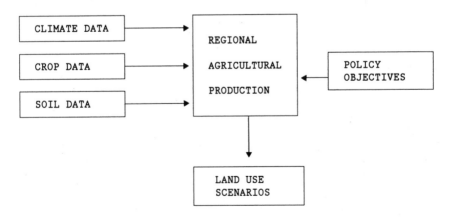

FIG. 4.   The framework of research needed to assess the relative contribution of climatic change to available options for agriculture.

grow under circumstances where water or nutrients limit growth and production. Under such circumstances the effects of climate change on agro-ecosystems will be much less prominent. Moreover, the possible effects of climate change on agriculture are outweighed by changes brought about by shifting policies. Stimulating agricultural production can easily lead to a doubling of output so that any effects of climate change would be overshadowed. Nevertheless, they are present and may alter the range of options available. An analytic methodology to evaluate the possible options for future land use and to demonstrate the consequences of various policies has been described. This methodology allows assessment of the relative importance of various changes and generates a set of feasible options for land use; the consequences of climate change for such options can be demonstrated.

   We propose the following research agenda to improve the quality of such studies.

Climate change. There is an urgent need for consistent and complete pictures of climate change. Not only the certain effects, such as CO_2 enrichment, but also the uncertain effects on temperature, precipitation and radiation in various agro-ecological zones should be included in such studies. There should be a regional breakdown of the results, ideally to the level of agro-climatic regions.

Impact assessment. The consequences of environmental change on basic physiological and phenological processes of plants, crops, pests and diseases should be quantified in more detail. This can be done through detailed studies under controlled conditions. Methods should be developed to study long-term effects because these may differ considerably from short-term effects.

Scenario studies. Studies on a higher integrative level, using scenario approaches, should be applied to explore various options for land use under changed climate conditions.

References

Campbell WJ, Allen LH, Bowes G 1988 Effects of CO_2 concentration on Rubisco activity, amount, and photosynthesis in soybean leaves. Plant Physiol 88:1310–1316
Cure JD, Acock B 1986 Crop responses to carbon dioxide doubling: a literature survey. Agric For Meteorol 38:127–145
Ehret DL, Jolliffe PL 1985 Photosynthetic carbon dioxide exchange of bean plants grown at elevated carbon dioxide concentrations. Can J Bot 63:2026–2030
FAO 1987 FAO production yearbook. FAO, Rome
Farquhar DG, von Caemmerer S 1983 Encyclopedia of plant physiology, vol 12B: Modelling of photosynthetic response to environmental conditions. Springer-Verlag, New York
Gifford RM 1979 Growth and yield of carbon dioxide-enriched wheat under water-limited conditions. Aust J Plant Physiol 6:367–378

Gifford RM, Morison JIL 1985 Photosynthesis, water use and growth of a C-4 grass stand at high CO_2 concentration. Photosynth Res 7:77–90

Goudriaan J 1990 Primary productivity and CO_2. In: Goudriaan J, van Keulen H, van Laar HH (eds) The greenhouse effect and primary productivity in European agroecosystems. Pudoc, Centre for Agricultural Publishing and Documentation, Wageningen, p 23–26

Goudriaan J, Bijlsma RJ 1987 Effect of CO_2 enrichment on growth of faba beans to two levels of water supply. Neth J Agric Sci 35:189–191

Goudriaan J, de Ruiter HE 1983 Plant growth in response to CO_2 enrichment, at two levels of nitrogen and phosphorus supply. Neth J Agric Sci 31:157–169

Goudriaan J, Unsworth MH 1990 Implications of increasing carbon dioxide and climate change for agricultural productivity and water resources. In: Kimball BA (ed) Impact of carbon dioxide, trace gases and climate change on global agriculture. American Society of Agronomy, Madison, WI, p 111–130 (Spec Publ 53)

Habjorg A 1990 Effects of climatic changes on growth and development of North European plants. In: Goudriaan J, van Keulen H, van Laar HH (eds) The greenhouse effect and primary productivity in European agroecosystems. Pudoc, Centre for Agricultural Publishing and Documentation, Wageningen, p 11–16

Kimball BA 1983 Carbon dioxide and agricultural yield: an assemblage and analysis of 430 prior observations. Agron J 75:779–788

Kimball BA, Mauney JR, Radin JW et al 1986 Effects of increasing atmospheric CO_2 on the growth, water relations, and physiology of plants grown under optimal and limiting levels of water and nitrogen. In: Response of vegetation to carbon dioxide. US Department of Energy and US Department of Agriculture, Washington, DC

Lemon ER (ed) 1983 CO_2 and plants: the response of plants to rising levels of atmospheric CO_2. Westview Press, Boulder, CO

Lenssen GM 1986 De invloed van CO_2 in combinatie met fosfaatgebrek op de groei en de transpiratie van tarwe. MSc thesis, Wageningen Agricultural University, Wageningen

Mortensen LM 1983 Growth response of some greenhouse plants to environment. X. Long-term effect of CO_2 enrichment on photosynthesis, photorespiration, carbohydrate content and growth of Chrysanthemum morifolium Ramat. Agricultural University of Norway, As (Sci Rep 272, vol 62)

Netherlands Scientific Council for Government Policy 1992 Ground for choices: four perspectives for the rural areas in the European Community. Sdu Publishers, The Hague (Rep Gov 42)

Nonhebel S 1990 The impact of changes in weather and CO_2 concentration on spring wheat yields in western Europe. In: Goudriaan J, van Keulen H, van Laar HH (eds) The greenhouse effect and primary productivity in European agroecosystems. Pudoc, Centre for Agricultural Publishing and Documentation, Wageningen, p 48–51

Norby RJ, O'Neill EG, Luxmoore RJ 1986 Effects of atmospheric CO_2 enrichment on the growth and mineral nutrition of Quercus alba seedlings in nutrient-poor soil. Plant Physiol (Bethesda) 82:83–89

Parry ML 1990 Climate change and world agriculture. Earthscan Publications, London

Rabbinge R 1986 The bridge function of crop ecology. Neth J Agric Sci 34:239–251

Rabbinge R, van Latesteijn HC 1992 Long-term options for land use in the European Community. Agric Syst 40:195–210

Rosenberg NJ 1981 The increasing CO_2 concentration in the atmosphere and its implication on agricultural productivity. Clim Change 3:265–279

Rosenberg NJ, Kimball BA, Martin PH, Cooper CF 1990 From climate and CO_2 enrichment to evapotranspiration. In: Waggoner PE (ed) Climate change and US water resources. Wiley, New York, p 151–175

Schlesinger ME, Mitchell JFB 1985 Model projections of the equilibrium climatic response to increased carbon dioxide. In: MacCracken MC, Luther FM (eds) The potential climatic effects of increasing carbon dioxide. National Technical Information Service, Springfield, VA (DOE/ER0237, US Dep Energy, Wash, DC) p 81–147

Schwarz M, Gale J 1984 Growth response to salinity at high levels of carbon dioxide. J Exp Bot 35:193–196

Sionit N, Hellmers N, Strain BR 1980 Growth and yield of wheat under CO_2 enrichment and water stress. Crop Sci 20:456–458

Squire GR, Unsworth MH 1988 Effects of CO_2 and climatic change on agriculture, 1988. Report to the UK Department of Environment, University of Nottingham, Nottingham

Strain BR, Cure JD 1985 Direct effects of increasing carbon dioxide on vegetation. National Technical Information Service, Springfield, VA (DOE/ER0238, US Dep Energy, Wash, DC)

Valle R, Mishoe JW, Campbell WJ, Jones JW, Allen LH 1985 Photosynthetic responses of 'Bragg' soybean leaves adapted to different CO_2 environments. Crop Sci 25:333–339

van Diepen CA, van Keulen H, Penning de Vries FWT, Noy IGAM, Goudriaan J 1987 Simulated variability of wheat and rice yields in current weather conditions and in future weather when ambient CO_2 has doubled. CABO, Wageningen (Intern Rep Centre Agrobiol Res Dep Theor Prod Ecol 14)

van Latesteijn HC 1993 A methodological framework to explore long-term options for land use. In: Penning de Vries FWT, Teng PS, Metselaar K (eds) Systems approaches for agricultural development. Kluwer, The Hague, p 445–455

Wilson CA, Mitchell JFB 1987 A doubled CO_2 climate sensitivity experiment with a global climate model including a simple ocean. J Geophys Res A 92(D11):3315–3343

Wong SC 1979 Elevated atmospheric partial pressure of CO_2 and plant growth. I. Interactions of nitrogen nutrition and photosynthesis in C_3 and C_4 plants. Oecologia 44:68–74

DISCUSSION

Sutherst: Are you making the assumption that the variability in climate will be unchanged in your climate change scenario?

Rabbinge: Yes, because we do not have any indication that the stability of the climate will decrease. We are certain that the CO_2 concentration will increase, but are uncertain about the temperature increase. Therefore, we have taken into account the CO_2 increase and the temperature increase separately and in combination.

Sutherst: So you are working with averages all the time.

Rabbinge: Yes, but we have also considered the variation in temperature and the consequences of that.

Oeschger: From discussions with biologists, I have the impression that although an increase of growth is initially expected or seen in response to an increased CO_2 level, a saturation or even a deterioration might follow. Also, I showed (p 15) that deconvolution of the observed CO_2 increase indicates only a minor non-fossil sink or source up to the present.

Rabbinge: You're right that the long-term effects are not considered, but that's something we have little experience with. There may be adaptation to the higher carbon dioxide concentration, and that may affect other processes, physiological processes, but we are not yet sure of that. More experimental evidence is needed to confirm that the long-term effects are different from the immediate effects. The immediate effects result from the basic processes, which can be exactly quantified and can be integrated into simulation models. With these models, you can evaluate the crop's response to climate change.

Mansfield: Do stomatal responses to CO_2 and their effects on water use vary between different species or different cultivars?

Rabbinge: Within some plant groups there are different types of stomatal behaviour, but normally the differences are between groups of plants. Sunflowers, for example, are a typical example of a group of plants which always have open stomata, whereas the cucumber or the soya bean have regulated stomata. In wheat the stomatal regulation is such that the internal and external carbon dioxide concentrations show a fixed ratio.

Mansfield: It would be possible to ask plant breeders to produce cultivars with different stomatal sensitivities. You could opt to have the same yield as at present, but with a better efficiency of water use, by growing a plant with a large stomatal response to CO_2.

Rabbinge: That's right. Some plants in fact behave like that, which means that they profit in terms of efficiency of water use when carbon dioxide is higher.

Mansfield: That might enable you to grow crops in areas that are insufficiently supplied with water unless they are irrigated.

Rabbinge: Or you could reduce the amount of irrigation.

Elliott: You discussed the relationship between CO_2 levels and yield, and made the point that this was an ecological correlation and not proof of causality. This is a problem that we face in epidemiology all the time. We often don't have the advantage that you have with crops, that of experimental design. You spend a lot of time on simulation, but rather than relying on simulation, you could actually set up experiments to test these things, which often we can't do in epidemiology.

Rabbinge: What we can do is test the reliability of the models we are using for the feasibility studies in which we use simulation models. The models have been tested at the three levels. They simulate experiments which were done under well-known conditions and the environmental conditions were incorporated into the models. These models gave reliable results, so we decided that they could be used for feasibility studies on climate change, provided that the response of the stomata is like that observed in the detailed experiments. We are therefore confident that the outcomes of the simulations are reliable. On this basis, you can also see what the economic consequences are of a carbon dioxide increase.

Hoffmann: Are there any assumptions built into your studies about the use of pesticides to control certain pests, to allow maximum productivity, or are the models independent of pesticide usage?

Rabbinge: At the level of individual fields, where we assume the conditions are optimal, the assumption is that pests and diseases are eliminated. This could be done in three different ways, through growing resistant varieties, using biological control measures or using pesticides. At a higher aggregation level, the European Community, land use could be considerably lower than at present; if land use is less, much smaller amounts of pesticides and nitrogen would be used than at present, and the environmental side effects would be decreased. We would be producing on better soils at a higher production level, and would be using external resources more efficiently. This seems to be a little counter-intuitive, because we always have the idea that with an increase of a growth-stimulating factor, the law of diminishing returns would apply. That's true if you consider only one growth-stimulating factor if all the others are abundant. In agriculture, it's the combination of factors that is important. You should not irrigate if you are not giving nitrogen, and you should not over-use nitrogen if you are not irrigating. If the combination is right, the efficiency of use of each of the resources does not decrease, but increases, or stays more or less the same, when yields perhaps increase. At the higher aggregation level, we have included the concept of 'best technical means', which means that usage of each of the external resources, nitrogen, phosphorous, pesticides or whatever, is the minimum per unit of product, at the same time maximizing the efficiency of use of all the other resources. This system can be developed and applied, and will result in high yields with very low input.

Lake: In the light of the predicted changes we heard about from Professor Oeschger, do you envisage large changes in the nature of land use, or is the existing infrastructure so conservative that we shall stay with the current portfolio of crops but grow them in different ways?

Rabbinge: I would think that the environmental changes, increases in temperature and carbon dioxide, will have less impact on land use than the socio-economic changes. For example, up to 1958, the countries that formed the European Community imported most of their sugar from other parts of the world, whereas nowadays the EC exports sugar. With the founding of the EC, a fixed, sustained high price for the sugar was guaranteed. If the EC now moves in a more market-orientated direction, we would probably begin to import sugar, and the sugar beet industry would suffer. This would be an example of socio-economically determined change. The tax-payer might no longer be willing to give so much money to a farming industry which is working in a less efficient way than it could be. That would cause more dramatic shifts in land use, in my opinion, than environmental changes.

James: Your comments have made me understand a little better the basis of the enhanced efficiency you describe. I am still not clear whether you are dealing with a *theoretical* series of assumptions about synergism and efficiency, or whether you have actually shown this in practice. If you have shown it in practice, presumably the extraordinary savings in nitrogen would also apply to phosphate, etc. How do the calculations eventually come out in terms of energy use? My predecessor, Sir Kenneth Blaxter, frequently used to point out that in fact Western agriculture is an *incredibly* inefficient industry in relation to energy use. Essentially, energy in the form of fossil fuel is being poured into the land for a minute return.

Rabbinge: What is important here is the added energy—machinery, pesticides, fertilizers, etc. If you include the added energy in the computation, the efficiency is higher when yields are high. Being thrifty with external input in a good production situation in fact decreases efficiency rather than increasing it. One should aim to use better soils, high production levels, and low external resources per unit of product tailored to the specific needs of the crop; then you will achieve higher efficiency than in a situation in which there is a lot of spoilage of external resources.

James: A net input is still needed.

Rabbinge: A net input is always needed. If you go back to a situation without inputs, you would be back to the situation that existed more than a century ago, where harvests were not much more than 1500 kg wheat/ha. To achieve this yield without machinery, you have to use manpower instead, and the resulting efficiency would be even lower. About 370 hours of labour would be needed per hectare of wheat, whereas nowadays in the United States the figure is 6 h/ha and in the United Kingdom 15 h/ha. In Western Europe today wheat yields are 7500 kg/ha, whereas at the beginning of the century the yield was 1500 kg/ha. This means that labour productivity has risen 200-fold over a period of about 80 years.

Oeschger: In Third World countries more than half of the world's population is suffering from famine. Global change should help to improve the living conditions of people from the Third World. How can the techniques you described help people in developing countries now?

Rabbinge: Many techniques are being implemented in many places already. In South-east Asia, and also in India and China, there is rapid development at present, but Africa is far behind. There is not much hope that the dramatic changes which are urgently needed in Africa will happen within 40 or 50 years. The second green revolution took place in Indonesia, India and China. In Indonesia over the last 10 years there has been an average yield increase of 150 kg/ha/year. The figures in India and China are similar. This change is needed in view of the growing populations of these countries. In Africa and in some parts of western Asia there is still an urgent need for a green revolution. An enormous input of technology is not needed; what is required is concerted action to adapt and improve the various production systems to enable these countries to buy external inputs. There is more agricultural research going on in Africa than anywhere else, but the results are not successfully implemented, because the farmers are not yet familiar with the technology and do not possess the means to get external resources.

Food quality and human nutrition

W. P. T. James

Rowett Research Institute, Greenburn Road, Bucksburn, Aberdeen AB2 9SB, UK

Abstract. New nutritional analyses suggest that current trends in the production of food are inappropriate for the health of most of the world's populations. Four deficiency problems now dominate analyses of the nutritional disorders of developing countries: the risks from iodine, vitamin A and iron deficiencies and protein energy malnutrition now affect over two billion children and adults. Chronic energy deficiency affects half of Indian adults, with similar rates in Pakistan and Ethiopia. India will need to increase food production two- to three-fold by 2020 to cope with the predicted population explosion and desirable increases in food consumption. As erosion, salination and environmental degradation further limit land availability, current problems will overwhelm agricultural demand.

Societies increase their meat, milk and fat consumption as they become affluent, and suffer from heart disease, diabetes, obesity, cancers and a variety of other 'Western' public health problems. Agricultural production is then regeared inappropriately. The Second World has an agriculture system geared to 1940s Western concepts of high animal production. Russia now vies with Scotland and Northern Ireland for the highest heart disease rates in the world and has the fattest adults in Europe. Most major non-infective public health issues throughout the world are nutritionally related. Global warming will exacerbate these problems, but effective dietary change with less animal production could release land which could be used more efficiently.

1993 Environmental change and human health. Wiley, Chichester (Ciba Foundation Symposium 175) p 80–103

The issue of food quality and human nutrition is once more back on the international agenda after several decades of neglect. The purpose of this paper is to highlight global problems relating to food supply and health. Current analyses suggest that far too few assessments have been made of the links between agricultural production and human health. The impact of global warming seems likely only to exacerbate pre-existing problems. A few assumptions will be made based on concepts set out by the World Health Organization (WHO 1992) and by the Greenpeace report (Leggett 1990). What will not be considered from the WHO report, however, is the issue of contamination of food by microbes (food poisoning), toxic chemicals or pesticides. These are important issues, particularly in developing countries, but the assumption that food quality relates only to these problems is one of the misconceptions of the modern era. We have become so used to considering food safety in toxicological or microbial terms that we forget that this selective view of food quality stemmed from a post-Second World War belief that the desirable food pattern for maintaining nutritional health had been established. The only issue at that time was how best to increase agricultural production of toxicant-free and microbe-free food to meet all possible nutritional needs at the lowest possible cost. Meat and milk production were established as these were considered to be high quality foods guaranteed to help small children grow and thrive. The post-war policies which emerged in Britain, North America and Western Europe have been copied world-wide. This agricultural revolution has distorted traditional farming practices. Continued inappropriate education and advice and governmental subsidies have resulted in food patterns which now constitute a public health problem of immense proportions not only in the Western World and the Second World of Eastern Europe, but also in the affluent parts of the Third World.

Despite this, there remain huge problems of nutritional deficiencies which could be exacerbated by global warming. Solution of these problems will require remarkable changes in developmental strategies with changes in public policies, dietary patterns and water supplies as well as in agricultural production and processing.

Nutritional deficiencies in the Third World

There are three well-accepted major deficiencies in the Third World relating to food quality: iodine, vitamin A and iron deficiencies (WHO 1992).

Iodine deficiency

Of these three deficiencies, iodine deficiency is one of the most devastating, because it leads to a range of disorders which threaten over a billion (1000 million) people in Europe, Asia, Africa and South America (Hetzel 1989). In Asia 90% of the population of Bangladesh are at risk of iodine deficiency and this is serious enough to lead to goitre in one third of the population. In India there is also a high prevalence of goitre in the Himalayan/sub-Himalayan region, reflecting the geophysical basis of this disorder. The constant leaching of iodine from rocks during heavy rain and the washing of nutrients from the soils during monsoons and floods have depleted iodine in huge areas of India, Pakistan, Nepal, Bangladesh and China. The iodine content of the diet falls as plant sources become depleted of iodine. There are an estimated three million cretins in the world, with far greater numbers of children suffering from poor mental and physical development which is unresponsive to iodine therapy or prophylaxis, unless this is started early enough (United Nations 1987, 1989). Similar problems occur in huge tracts of Africa and South America (Table 1).

TABLE 1 Some estimates of the numbers of people (in millions) world-wide affected by specific nutritional deficiencies

Region	Vitamin A		Iron			Iodine		
	Blind or eye damage	Deficiency	Children	Men	Women	Cretinism	Goitre	At risk
Africa	1	5	95	23	47	0.50	30	60
South-east Asia	1	15	267	130	191	1.50	100	280
Other Asian countries	0.5	5	—[a]	—[a]	—[a]	0.90	30	400
Latin America	0.5	5	32	13	15	0.25	30	60
Total	3	30	394[a]	166[a]	253[a]	3.15	190	800

[a]China excluded.
Data compiled from United Nations (1987) with some estimation of regional values from totals for vitamin A.

Permanent major prophylactic programmes will need to be established to eliminate this huge global problem. If global warming leads to major increases in rainfall, then presumably new areas of depletion will arise over centuries as rocks become leached of iodine and the metabolically related element selenium (Arthur 1991).

Vitamin A deficiency

Vitamin A deficiency is another huge problem affecting both children and adults. It leads to blindness and an increased susceptibility to infections and premature death. There is still considerable controversy as to the best methods of preventing the problem (Gopalan 1992a). Claims that megadose therapy by injection is the most effective approach are countered by the recognition that in practice an injection can lead to cross-infection; the real need is for effective community education, to increase people's intake of the green vegetables and fruit such as mangoes from which vitamin A or its carotene precursors can be derived. This problem occurs in societies where green vegetables are often available, so food production seems to be a less important factor. Global warming, in theory, would exacerbate the problem if the water demands of vegetable growing become too great. What is not in doubt is the extent of the current and potential problem (Table 1).

Iron deficiency

Iron deficiency is one of the commonest deficiencies in the world. It leads to the development of anaemia with lassitude, and consequently lower productivity. Perhaps even more important is the emerging recognition that children need iron for brain development. There seems to be a specific period in infancy and early childhood when the development of the brain depends on an adequate supply of iron for the activity of iron-containing enzymes (Soewondo et al 1989). Experiments on animals suggest the availability of iron at a later stage does not rectify the deficit.

The important question is why children and adults develop iron deficiency. In general terms there are two principal factors: loss of iron because of intestinal parasitism and poor absorption of iron from a diet which is predominantly vegetarian.

These crude generalizations suggest two strategies, both of which might be affected by a global temperature rise. One way of reducing parasitism is to develop, as a major priority, better water management systems. Improved hygiene with a clean water supply and proper sanitation would make a big difference, but this is needed in rural as well as urban communities. A second option is to increase the meat intake of children and adults. Meat substantially increases the availability of inorganic iron derived from plant sources (by a mechanism involving the release of amino acids in the small intestine) and itself contains highly available iron in the haem form. If public hygiene in the Third World were near Western standards there would be no intrinsic need for an omnivorous diet, but where there are recurrent infections and parasitism the inclusion of some haem iron in the diet does reduce the prevalence of anaemia. However, even the provision of 20% of dietary iron in the haem form demands a substantial animal industry which would probably require land which could be used more efficiently for grain production. Any change resulting from global warming leading to an increase in intestinal parasitism is likely to increase the prevalence of iron loss and anaemia (Table 1).

Protein and energy malnutrition

Malnourishment in children is often defined in terms of both stunted growth and weight deficiency. New global analyses suggest that although the proportion of children in the population who are malnourished may be falling, the rate of population growth is increasing so that the absolute numbers of malnourished children are likely to grow, not decline (Table 2). New evidence is emerging to suggest that about 40% of the developing world's children show a sufficiently slow rate of growth to be classified as having a low height for their age. This degree of stunting is associated with a failure to develop mentally. A combination of food supplements and play therapy can have a substantial impact on both growth and mental development. The improvements resulting from only a few months intervention may be sustained for several years afterwards (Grantham-McGregor 1992). The links, if any, between the spurt in height and in mental development are uncertain, but there is accumulating evidence that the supply of protein, particularly animal protein, is an important determinant of longitudinal growth (Golden 1988). The inflow of protein, or indeed of specific amino acids, may induce an anabolic drive which stimulates bone growth. Protein derived from meat would also provide a range of nutrients, such as iron, zinc and selenium, in higher amounts or in a more bioavailable form, so protein intake in itself may have little to do with any effect on brain development associated with the increase in height. Clearly, these are complex processes, but for the purposes of the present discussion, the theme is once more that some benefit would be derived from eating modest amounts of animal protein.

It should be emphasized that, as with anaemia, if a child or adult is living in a hygienic environment, excellent development is achievable on a vegetarian diet. The issue becomes a practical one of assessing the relative costs and effectiveness of water and sanitation measures against those incurred in attempting to encourage some meat eating in children and adults who have no religious objections.

TABLE 2 Estimated prevalence and numbers of wasted and stunted preschool children in developing countries[a]

	Asia (excluding China)		Africa		The Americas		Total	
	Percentage	*Millions*	*Percentage*	*Millions*	*Percentage*	*Millions*	*Percentage*	*Millions*
Wasted[b]	16	33	7	4	4	2	12	39
Stunted[b]	40	81	35	24	43	20	39	125

[a]Data from WHO 1983, see Keller (1988).
[b]Percentage of children aged 0–4 years below the mean minus two standard deviations of the NCHS (National Child Health Survey) reference proposed by WHO.
Taken from Keller (1988).

Adult chronic energy deficiency

Some aid programmes are organized to provide food supplements to families specified as suffering from chronic energy deficiency (CED). This condition was undefined, so we proposed a scheme based on body weight and physical activity (James et al 1988) which we later refined so that it is now based on body weight in relation to height alone (Ferro-Luzzi et al 1992) (Table 3). We have undertaken a series of assessments throughout the world to validate its use (Shetty & James 1993). We found that underweight (by our definition) men and women in Asia and Africa tend to be sick, off work and less able to undertake strenuous work, and they have a reduced work capacity and lower productivity. In India a positive correlation was found between the degree of weight deficiency and the risk of death. Underweight mothers are also particularly liable to bear underweight babies who are at risk of death in infancy or poor long-term growth and mental development. This has serious implications for the 'human capital' of any society if the condition is common.

To our dismay, we found that half the adult population of India is malnourished, with about 10% having the severe, third degree form of CED. Table 3 shows some preliminary figures from different countries. These data suggest that we have misunderstood the nature of the global food problem. It is in India and its surrounding countries where the major problem exists, whereas

TABLE 3   Chronic energy deficiency (CED)[a] in young adults aged 18–29 years

	Men		Women	
	Total (% Population)	Grade III (% Population)	Total (% Population)	Grade III (% Population)
Asia				
China	11	1	13	1
India	50	9	49	11
Malaysia	11	—	15	—
Pakistan	48	—	40	—
Vietnam	28	1	29	2
Africa				
Togo	7	0	10	0
Benin	41	4	48	8
Ethiopia	49	5	58	6
Zimbabwe	15	1	12	1
Tunisia	5	0	5	1
S. America				
Brazil	8	1	11	1

[a]CED is defined on the basis of the same cut-off points of the body mass index (weight kg/height^2 m^2) for men and women: Grade I = 17.0–18.4; Grade II = 16.0–16.9; and Grade III = < 16.0. Data were provided to the FAO by a large number of national agencies and experts. More extensive analyses will be published by the FAO (Shetty & James 1993).

China is in a remarkably good nutritional state. War-torn parts of Africa and countries with areas of drought also have a high prevalence of CED, but health problems in many African areas seem to relate as much to communicable diseases as to food supply.

Future food needs in the Third World

We are currently recalculating the future food needs of the world using United Nations' population projection figures and a recently developed system for specifying the energy needs of a population (James & Schofield 1990). One example serves to emphasize the startling conclusions (Fig. 1). Here, we have calculated the increments in average national food intake needed for the underweight Indian population to achieve a healthy size, weight and level of physical activity, and even eventually an increase in adult stature and the establishment of a Western style population structure without any increase in numbers. On this basis alone an estimated increase of 28% is necessary. If,

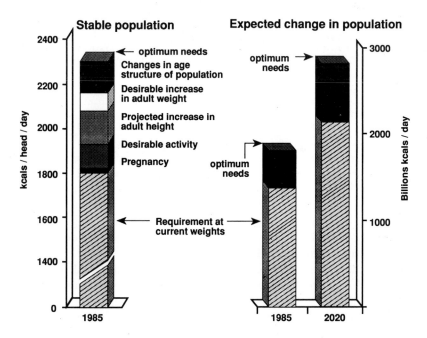

FIG. 1. Projected food needs of the Indian population. The individual requirements on the left (kcals/head/day) show the current requirements and estimated increases in energy intake necessary to achieve desirable changes in weight, height, activity and population structure. On the right, these requirements are multiplied for the whole population as of 1985 and for the population predicted for 2020, and are expressed in billions kcal/day.

however, we then estimate requirements for the projected population in the year 2020, the increase in food demand is startling (Fig. 1). Extensive analyses (James & Schofield 1990) show that population growth is the one factor which dominates all others in determining the tonnes of grain or its equivalents which will need to be grown.

This is a formidable enough challenge, but Gopalan, the eminent nutritionist and former head of the Indian Council of Medical Research and its Hyderabad laboratory, has assessed the problem of Indian food supplies as they affect Indian health. He considers that salination problems in irrigated areas, depletion of soil fertility through continued monoculture, mismanagement and erosion are threatening a pandemic of deficiencies—in the soil, leading to poor plant growth with deficient plants, and shortly in the population (Gopalan 1992b). It is estimated (FAO and Anonymous 1991) that the total land available in the developing world for agriculture is 780 million hectares and that there may be an additional 70 million hectares for cultivation. However, within 10 years there will be a loss of 70 million hectares through erosion and 82 million hectares through salination of irrigated areas. It is noteworthy that perhaps 35% of the world's food supplies are obtained from the 15% of land which is irrigated. This intensively cultivated land is in a state of decline. On this basis we can expect major food problems of quality as well as quantity, irrespective of any damaging effects or additional loss of land resulting from global warming.

Western and European agriculture and health

For decades, Western governments have assumed that their agricultural and food policies were well on course to provide the variety of safe foods which people demand. Pesticide residues were checked, toxicants were excluded or restricted and hygiene regulations and refrigerated transport and storage limited microbial contamination. Food became ever cheaper to buy, because of sustained subsidies for farmers, the provision of remarkable resources for agricultural research and advisory services and the protection of marketing cartels. Huge vested interests in the food and agriculture business therefore developed, and those concerned unsurprisingly find it impossible and indeed threatening to accept that the whole concept of food quality has been misplaced. Toxicological rather than nutritional thinking has dominated all public perceptions and policies to the detriment of millions of people who now suffer the effects of an array of chronic diseases, such as heart disease, high blood pressure, obesity, diabetes and cancers, which have a substantial nutritional basis.

In retrospect, we can see that the key error was to promote the unlimited consumption of animal products, fats and even sugar on the grounds that the dominant problems of society were energy and protein deficiencies. Although any specific energy or protein deficiency (for which there was no evidence) has been dealt with, we are now faced with hopelessly inappropriate food patterns, a misuse of large tracts of land for the most extraordinarily intensive livestock industry geared to producing meat and milk with an efficiency which, in short-term limited economic analyses, is astonishing. In practice, as Rifkin (1992) notes, ten-fold more water is required for producing butter or meat rather than cereals, and feeding cereals to cattle rather than chickens involves a four-fold loss of efficiency. Preston, from my own institute, developed the barley beef system which transformed intensive livestock rearing and led to the vast demand for cereals for cattle feeding throughout the world. It was a logical step to assess pig and then chicken conversion rates (weight of cereal protein to weight of meat protein) on cereal diets and soon the productivity gains in meat production were immense. Meanwhile, Eastern European governments, with their command economies, followed the perceptions of the day and organized a huge animal husbandry system to cater for the people's needs. Soon cereals were at a premium and the Soviet Union had to purchase grain from the United States or on the world market where prices fluctuated depending on the Russian harvests and the outcome of US–Soviet negotiations influenced by spying and satellite surveillance of harvest yields.

The concentration on meat and milk production has led to an appalling diet in Eastern Europe, with fat intakes of 45–50%, a saturated fatty acid intake of up to 30% and an escalating rate of heart disease which will shortly outstrip even the rates in Scotland and Northern Ireland. The percentages of overweight and obese adults in Eastern Europe are very high and exceed rates found anywhere else in the world except in odd spots such as the Pacific Islands or in the Pima Indians of Phoenix, Arizona.

Certain elements of the food industry now operate with the same ruthlessness as the tobacco industry (Taylor 1984), subverting any health-related developments in policy-making or education which might threaten their markets. Huge investments are made in Third World countries to exploit opening markets for sugar/cola drinks and for special meat products such as hamburgers and chicken. These are all marketed, as with cigarettes, by association with a portrayal of the affluent Western lifestyle. We therefore have an international food and agriculture business with a pervasive influence on government and international politics nearly as great as that of the military–industrial complex!

Table 4 presents new WHO goals for ideal nutrient intake for maintaining health, together with a demonstration that these match the views of other national bodies (Cannon 1992). The Chinese and Japanese, with their low fat diets based on rice and vegetables, are now overcoming their deficiency problems; the Japanese currently have the longest life expectancy in the world, but intensive advertising is promoting market penetration by Western food firms. Japanese fat intake has already risen to 28%. Coronary heart disease is now a worthwhile condition for young Japanese doctors to study because it has become a frequent problem in the main cities. Similarly, in Malaysia and Thailand, where there have been astonishing increases in gross national product,

TABLE 4 The WHO 1990 recommended population nutrient goals and other national goals

Nutrient	WHO Study Group's population nutrient goal		Other national upper or lower limit recommendation[a]		
	Lower limit	Upper limit	Lowest	Highest	Median
Total fat (% energy)	15	30	15	35	<30
SFA (% energy)	0	10	8	15	<10
PUFA (% energy)	3	7	10	13	10
P:S ratio[b]	—	—	0.45	1.2	1.0
Cholesterol (mg/day)	0	300	225	300	<300
Total CHO (% energy)[c]	55	75	40	70	55
Complex CHO (% energy)	50	70	—	—	—
Dietary fibre (g/day)					
As NSP	16	24	—	—	—
As total dietary fibre	27	40	20	35	30
Free sugars[d] (% energy)	0	10	9	25	<10
Salt (g/day)	—[e]	6	5	10	7

[a]Values for total fat, SFA, cholesterol, sugar, and salt are upper limits (the median value therefore implies that total fat intake, for example, should be less than 30% of energy); other nutrient values are lower limits (the median value therefore implies that PUFA intake, for example, should be greater than 10% of energy).
[b]Earlier reports specified P:S ratios. Later reports offered recommendations on classes of fatty acids but not on ratios. In this report, care is taken not to specify a ratio, because the SFA value could in theory be reduced to 0%, but the WHO Study Group recommends that PUFA intake should remain between 3% and 7%.
[c]Most reports suggest that most of the carbohydrate should be complex carbohydrate, without specifying a specific proportion.
[d]These sugars include monosaccharides, disaccharides and other short-chain sugars produced by refining carbohydrates.
[e]Not defined.
Taken from WHO (1990).
SFA, saturated fatty acids; PUFA, polyunsaturated fatty acids; P:S, ratio of polyunsaturated to saturated fatty acids; CHO, carbohydrate; NSP, non-starch polysaccharides.

the young city-dweller is changing to a Western diet and meat consumption is rising rapidly, as is the prevalence of obesity. The risks of heart disease are apparent—increasing cholesterol and blood pressure levels and rising smoking rates.

On a global basis, we can therefore see that we are moving to an extraordinary state of affairs. Perhaps a billion people are malnourished in the Indian subcontinent and neighbouring countries, with an escalating population placing impossible demands on the agricultural capacity and the threatened and diminishing soil fertility of the region. War in the Horn of Africa and drought in the Sahel and Zimbabwe are leading to states of desperation, but elsewhere in Africa there seems to be sufficient agricultural capacity despite the worrying fertility rates and almost total government inaction on population control. Yet in Europe and North America we squander huge resources and deplete the land with inappropriate policies geared to producing food unsuitable for a healthy diet.

It seems unlikely that Europe or the Americas will run into problems of food shortage or prove incapable of modifying food quality once consumer movements overcome the duplicity and vested interests of the food and agriculture industries, but the global problem is different. How can we begin to cope with the needs of Bangladesh, Nepal, Ethiopia and Somalia and anticipate the needs of the Indian subcontinent as a whole? There will need to be a continual transfer of food from Europe and North America to Asia and parts of Africa over many decades. The likely lack of political leadership means that revolutionary changes in world financial and organizational strategies will be needed for this to happen. The more likely scenario is that famines will persist, and there will be rising rather than falling rates of infantile and childhood death as voluntary organizations struggle to cope with the huge burdens. Governments, as now, will transfer minimal funds while expressing great concern. Global warming will simply exacerbate these trends and disparities in both food quality and quantity. As Western societies are educated in the unrealistic philosophy, and therefore spurious policy, of individual free choice, the demand for national and international action will slowly decline. Western consumerism will therefore dominate our perspectives on food quality unless environmentalists become more effective lobbyists. There seems little prospect of coping with the global issues unless the United Nations becomes a much more powerful force for the good of humanity. This is where the battles lie.

References

Anonymous 1991 The green counter-revolution. Economist 319:107–108

Arthur JR 1991 The role of selenium in thyroid hormone metabolism. Can J Physiol Pharmacol 69:1648–1652

Cannon G 1992 Food and health: the experts agree. Consumer's Association, London

Ferro-Luzzi A, Sette S, Franklin M, James WPT 1992 A simplified approach of assessing adult chronic energy deficiency. Eur J Clin Nutr 46:173–186

Golden MHN 1988 The role of individual nutrient deficiencies in growth retardation of children as exemplified by zinc and protein. In: Waterlow JC (ed) Linear growth retardation in less developed countries. Raven Press, New York (Nestlé Nutr Workshop Ser 14) p 143–163

Gopalan C 1992a Vitamin A deficiency and childhood malnutrition. Lancet 340:177–178

Gopalan C 1992b Challenges and frontiers in nutrition in Asia. In: Proceedings of the 6th Asian congress on nutrition, p 1–20

Grantham-McGregor SM 1992 The effect of malnutrition on mental development. In: Waterlow JC (ed) Protein-energy malnutrition. Edward Arnold, London, p 344–360

Hetzel BS 1989 The story of iodine deficiency. Oxford University Press, Oxford

James WPT, Schofield EC 1990 Human energy requirements: a manual for planners and nutritionists. (Published by arrangement with the Food and Agriculture Organization of the United Nations) Oxford University Press, Oxford

James WPT, Ferro-Luzzi A, Waterlow JC 1988 Definition of chronic energy deficiency in adults. Report of a working party of the International Dietary Energy Consultative Group. Eur J Clin Nutr 42:969–981

Keller W 1988 The epidemiology of stunting. In: Waterlow JC (ed) Linear growth retardation in less developed countries. Raven Press, New York (Nestlé Nutr Workshop Ser 14) p 17–39

Leggett J (ed) 1990 Global warming. The Greenpeace report. Oxford University Press, Oxford

Rifkin J 1992 Bovine burden. Geographical 64:11–15

Shetty PS, James WPT 1993 Body mass index: an objective measure for the estimation of chronic energy deficiency in adults. Report to FAO, in press

Soewondo S, Husaini M, Pollitt E 1989 Effects of iron deficiency on attention and learning processes in preschool children: Bandung, Indonesia. Am J Clin Nutr 50:667–674

Taylor P 1984 Smoke ring: the politics of tobacco. Bodley Head, London

United Nations 1987 ACC/SCN (Administrative Committee on Coordination Subcommittee on Nutrition) First report on the world nutrition situation. United Nations, Geneva

United Nations 1989 ACC/SCN (Administrative Committee on Coordination Subcommittee on Nutrition) Update on the nutrition situation. United Nations, Geneva

WHO 1990 Diet, nutrition, and the prevention of chronic diseases. Report of a WHO Study Group. WHO, Geneva (Tech Rep Ser 797)

WHO 1992 Our planet, our health. Report of the WHO Commission on Health and Environment. WHO, Geneva

DISCUSSION

Elliott: You showed that mortality increases in relation to energy deficiency. In Western populations, for example in the studies of male Whitehall civil servants, the same phenomenon is observed. If you divide the, presumably, not underfed group into quintiles on the basis of body mass index, you find that the lowest quintile has a relatively high mortality, generating the so-called U-shaped or J-shaped curve (Jarrett et al 1982). How does this relate to your work at much lower body mass indices?

James: Are you asking me to explain why people die? We are currently doing a study on immunological competence.

Elliott: I'm saying that the same phenomenon is observed in Western populations who presumably aren't starving, although there were some people with very low BMIs in that group. The 20% of that population of Whitehall civil servants with low BMI had a higher mortality than those with a higher BMI.

James: There are three aspects to this. When people are sick, they lose lose weight, and therefore in epidemiological analyses you have to strip out the first 2–5 years of excess deaths in those with a low body weight to exclude low weights relating to illness. This needs to be distinguished from a low body weight which eventually predisposes people to becoming ill. Secondly, the question is whether or not you smoke, because smoking makes you thin, and it is a killer.

Elliott: The same phenomenon is found among smokers though. Figure 1 shows mortality rate per thousand person years by quintiles of body mass index in the Whitehall study. The rates are adjusted for age and grade of employment. Among smokers, of a total of 2255 deaths, the rate of mortality in the lowest quintile of body mass index ($\leqslant 22.4 \, \text{kg/m}^2$) was exceeded only by that of the highest quintile ($\leqslant 27.0 \, \text{kg/m}^2$). This ranking was unchanged when the first five years of follow-up was excluded, or when men with 'disease at entry' were

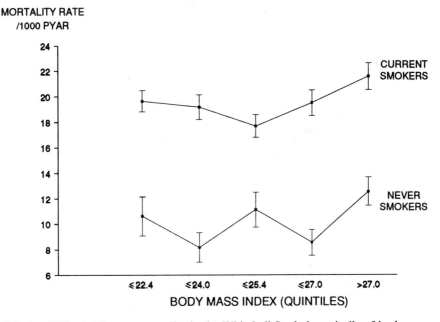

FIG. 1. (*Elliott*) All-cause mortality in the Whitehall Study by quintile of body mass index (kg/m²) and smoking habit. The rates of mortality, given as rate per 1000 person years, are adjusted for age and grade of employment. Error bars are SEM.

excluded, including those with unexplained weight loss during the previous year (Martin Shipley, personal communication).

James: The third aspect is that in studies in both children and the elderly, but not yet in adults, a whole series of tests reveal defects in immunological reactivity which can be rectified by feeding. My hunch is that the increased illness in undernutrition, and perhaps the increased mortality, relates to a decreased resistance to infection.

Elliott: There is a vicious circle here. If you are ill, your activity is low and you don't eat.

James: That's why I stressed that this is only a hypothesis.

Edwards: It has always, quite reasonably, been believed that it is extremely bad to be starved during pregnancy. However, the counter example, which has been beautifully documented, is that of starvation after the Second World War in The Netherlands, in Walcheven in particular. This cohort of babies has been followed through school and university and they are now in their late 40s. The good news was that after the appalling starvation during pregnancy their mothers had experienced, and the poor post-natal nutrition available, these children developed to be almost indistinguishable in terms of size and academic achievement from those born in happier times. The terrible position in India must be related to the chronic malnutrition which precedes the pregnancy.

James: One must distinguish between chronic and acute energy deficiency. Those Dutch data showed that amennorhoea was one of the first responses to weight loss as these women went into semi-starvation. The children that were conceived and born and eventually grew up tended to be plumper than average; I think that's because it was the more obese women who were able to maintain their menstrual cycling and become pregnant, and go through the pregnancy. That's a form of selection. The women who were semi-starved during the later part of pregnancy had babies weighing about 250 g less than normal. We are talking about a very substantial shift in the overall birth weights in India compared with the Dutch data. We must distinguish between an acute effect in otherwise fairly well-nourished Dutch women and this chronic sustained effect in India.

Rabbinge: You showed that a tremendous increase in food production is necessary in India because of the increase in population. You did not consider an alteration in diet. Did you assess what the situation would be with another diet? When income rises the diet would probably shift towards animal proteins, which would necessitate more plant production.

James: I did those calculations only in terms of energy. There's little evidence that cereal energy is used differently from fat energy; there is a difference of maybe 2 or 3%. People tend to overeat fat energy, and that's related to appetite control. The strategic problem, of course, is that you would be much better off eating the energy from cereals or from vegetable fat than feeding the plant foods to an animal and then consuming the animal. That is a very wasteful system.

Rabbinge: Between 7 and 10 kg cereal is needed for each kg of meat that is consumed.

Hautvast: Professor James, you began by discussing the extent of malnutrition in the world. The problem is that the three large multinational agencies, WHO, FAO and the World Bank, use different figures of world malnutrition. Is it the case that there are no good estimates of the prevalence of malnutrition? The body mass index is probably a very good index for this purpose.

James: The standard calculations of world malnutrition actually use inappropriate statistical calculations on household food availability and agricultural food balance sheets. These organizations then have to use figures which we gave them for the weight of children, the size of the adults, their energy needs and the population distributions and numbers before coming up with an estimate of malnutrition. These figures are indirect, and are all crude estimates. The value of doing something objective, such as measuring heights and weights, is that we can predict more accurately the final outcome in terms of the real effect on children and adults. It's simple and you don't have to spend five years juggling agricultural statistics to come up with a value which you can only hope is right. This technique is beginning to make people realize that some of the agricultural statistics are wrong. If you want to ask how many hundreds of millions of people are likely to come into this undernourished category, you can do some simple arithmetic, and will come up with a value of about three quarters of a billion adults world-wide.

Rabbinge: There has been a lot of debate on the decreasing rate of increase in rice yields in many countries in Asia, including India and the Philippines. This was thought to be due to zinc or other deficiencies. Over the last few years we have been trying to determine in the Philippines whether one reason could be that the nitrogen is not in the crop when it is needed. If you apply split nitrogen treatments with one treatment around flowering, the yield increases again. This has been experimentally demonstrated during this last season in four or five places in India and the Philippines. You may wonder why they did not do such experiments earlier. All the data on rice fertilization were from old experiments that did not take into account other growth-limiting factors and synergism between them. This meant that even with a marginal increase in input, such as a nitrogen treatment, especially at the later phases of development, a major increase in yields could be achieved. If you deal with the inputs in the proper way, higher yields result. This is all experimental, and has not yet been demonstrated in the farmer's field, so I have to be cautious. If the theoretical outcomes prove correct in an experimental setting, and if it's possible, at least in the better fields in the Punjab, for example, to get the same results, yields could continue to increase for some time. This may help to solved the problems not only in India but also in the rest of Asia. Only 18% of the world's cultivated land is situated in China, India and the other countries of Asia where more than 50% of the world's population lives and where population is still increasing.

Lake: Increasing carbon dioxide in the atmosphere increases, for some species, the ratio of roots to shoots, and it could be that the plants will explore more of the soil, so at least in the medium term, though clearly not in the longer term, they may be able to extract more micronutrients.

Another point I would like to make concerns the alarming increase in energy supply that will be required by this growing population to avoid malnutrition. To what extent will the increasing global carbon dioxide concentration be likely to increase food production, other agronomic factors being unchanged?

Mansfield: The effect on root to shoot ratios depends on the species. The ratio is more likely to increase with perennial than with annual crops, and most of the Indian crops we are talking about are annual pulses. One would have to deal with each crop separately to answer this question, but it's an important one and something we do need to know about.

Rabbinge: If the increase in carbon dioxide concentration is to lead to increased yields, all the other factors which limit yields and growth would have to be eliminated. As I indicated, calculated potential increases in yield resulting from enrichment of the environment with carbon dioxide are all low, the maximum being of the order of 10%. It is far from certain that there will be a considerable increase in potential yields, because there are data indicating that photorespiration will increase. The positive effects on photosynthesis and water use efficiency might in fact be counterbalanced by the increase in photorespiration.

Lake: Carbon dioxide certainly has a beneficial effect on glasshouse crops. Some commercial glasshouse crops are now never grown without additional carbon dioxide because it has huge economic benefit in terms of productivity. Under highly controlled, ideal conditions there is no doubt about the physiological consequences. Do models of the field situation give grounds for optimism? If the crops to feed the starving people in the developing world could be grown in greenhouses, a doubling of CO_2 concentration might feed a 50% increase in population, but you have suggested that the benefits might be only 2–3%.

Rabbinge: Even if you were an optimist, 10% would be as much as you could hope for.

Mansfield: Greenhouse crops tend to be CO_2-limited in the absence of additional CO_2, because the greenhouse excludes atmospheric CO_2. However, when all the data are taken into account, it appears that CO_2 provided under optimum conditions is a good thing.

Rabbinge: That only applies where CO_2 is the limiting factor. As soon as it's not the limiting factor, when nitrogen or other inputs such as phosphate become limiting, the effect is much less.

Zehnder: Professor James, you said that nutrition, energy input and a balanced diet are important. We have been talking about yield increases per hectare. Should the aim of agriculture be to provide maximum energy or balanced food production? What should the ratio be between animal and plant protein?

James: At one extreme is China, where about 19–20% of the total protein intake comes from animal sources, and at the other are some of the European countries, where the proportion is 70–75%. How much animal protein a person needs is a very interesting question. In theory, if you live in a wonderful society where you don't have problems with iron absorption and parasitism, you can be a successful vegetarian, or even a successful vegan if you give yourself vitamin B_{12} injections, and you don't need meat or animal protein at all. However, most of the major societies of the world actually derive some benefit from eating some meat. The ideal figure may be roughly 30% of total protein from animal sources, which is what is found in Thailand, for example, a country which has extraordinarily low rates of heart disease, obesity and cancer and rapidly increasing lifespan, associated with very low levels of deficiency diseases. We haven't actually locked into our agricultural policies for the future any nutritional health criteria, but these are necessary if we are to avoid huge economic problems in the health sector, let alone in the agricultural sector.

Zehnder: You say both quality and quantity need to be increased, and that ideally about 30% of the protein in the diet should come from meat. In Russia and in parts of Eastern Europe about 80% of protein intake comes from meat. It will be difficult to change this habit because it is part of the culture. Might the potential food shortages of the next century be limited by societies altering their habits and eating less meat?

James: For the WHO, we have actually derived a series of population-averaged nutrient goals from which individual needs or population needs can be calculated. The dilemma will be essentially political. If there is a huge demand for food in certain parts of Africa and Asia, where the environmental constraints are going to become appalling in relation to population pressures, and at the same time in Europe we enter a phase of super-efficient use of energy and nutrients for maximum food production, do you really think that Europeans will ship out to Asia, for ever, hundreds of millions of tonnes of food per year?

Zehnder: If money can be gained, that would probably work. I would be more concerned about what happens when the food arrives; a transport system is needed urgently within these countries.

Klein: Extensive research in Europe and the USA has more or less answered the question of what we should eat, the questions of food quality and health. However, what impact does this have on agricultural policies? These are two different issues. We know that we should eat no more than 25 g fat per day, and at least 25 g of fibre per day. The question is whether the food companies and the agricultural community are matching this need for healthy nutrition. Perhaps one problem is that we know what we should eat but we don't like it.

James: Western societies are in precisely that position. The current agricultural policies that you and I automatically assume are set in a contemporary context were actually set by Western governments during the Second World War and immediately afterwards. The feeding systems that were used in Germany and in Britain and Norway and everywhere else during the Second World War were spectacularly successful. Governments concluded that they now knew what human needs were, and the top priority for Britain and many other countries was to become self-sustaining in food, but with a major animal-production programme. Society (civil servants, economists and everybody else) is locked into agricultural policy that relates to the priorities that existed in the 1940s, and nobody has yet rethought agricultural policy to make it appropriate to our current understanding of public health. The usual response to this criticism is that the system is only providing what the customer wants. This is total nonsense. There's no such thing as the free market—it's highly geared, with huge subsidies, and, of course, a massive investment, such that there are now cartels. Take sugar beet, for example. There are very good economic reasons to grow sugar beet but this is totally unreasonable from the human health point of view. This is similar to the situation in the tobacco industry, which is keeping smoking going. We now have to try to re-set major national goals in the teeth of opposition from those with vested interests which are perfectly understandable, and are actually not without some merit, because they were based on important criteria from the 1940s.

Edwards: What proportion of the energy going into a cow comes out as meat or milk?

Rabbinge: To produce 1 kg of meat or dairy products, the cow needs to be fed between 7 and 10 kg of wheat (on a dry weight basis).

James: When I was at the WHO about 10 years ago, the Chinese head of public health was there, and he was horrified to hear this because they had just initiated a major pig production programme. The Chinese then decided that they should shift to chickens. The efficiency of chicken production is still only 20–25%. Major changes in the ecosystem in Chinese villages would be required to increase this type of animal protein production. It's actually the sheep, the goats and the cattle, the ruminants, the animals that can graze or browse on land that cannot be used by man, that are more likely to be the salvation from the animal point of view than intensively reared pigs and chickens. Although intensive animal production is more efficient nominally, it actually diverts food resources which could otherwise be used directly.

Rabbinge: If you relied on these scavenging, grazing animals, your diet would involve very low meat consumption.

James: I said that 30% of protein should be derived from animal sources, but I didn't say it should all be from meat—I was including milk and cheese as well.

Zehnder: We talked earlier about the effects of CO_2 and temperature on crop yields. We haven't really discussed how higher temperatures will enhance the activity of pests and hence lead to an increase pesticide use. When these chemicals get into the food, what effect will they have on health? Might the health benefits of increased productivity be counterbalanced by the adverse effects of pesticides?

Rabbinge: With increased temperature there will not necessarily be more pests and diseases, but there will be different pests. For example, if there is a temperature rise in western Europe, one of the most important pests of corn, the cornborer, *Ostrinia nubilalis,* which is at present no further north than Paris, would become a problem in The Netherlands. There are also the pest's natural enemies to consider. The question is whether you can manipulate the system with appropriate biological control. The use of pesticides could drop dramatically, for three reasons. First, crops should be grown in rotation, so that it is not necessary to spray or to fumigate as a preventative measure against cyst nematode damage. That would eliminate a lot of pesticide usage. In The Netherlands, for example, more than 50% of all pesticides are used for fumigation, which is not necessary if crops are grown in rotation over 5–6 year cycles. The second reason is that biological control will increase, and will be combined with pesticide use to achieve fully integrated pest management, which again will eliminate a lot of pesticides. Finally, nitrogen and other inputs necessary for plant growth should be applied only to the better soils and at the moment when the crop needs it. This would increase the efficiency of the use of external inputs considerably. For example, in The Netherlands, wheat grown in sandy areas has to be sprayed twice as often as in the clay areas, because there are more problems with mildew. Pesticide use could decrease if we worked in the better sites in a rational manner. Unfortunately, we are far from doing it that way.

Hoffmann: What is known about the effects of ozone on crop yields? Will the predicted general increase in tropospheric ozone to 75 parts per billion by volume from the current level of 20 or 30 have an impact on crop yields? Likewise, will increased UVB have a negative impact?

Mansfield: UVB is only just beginning to be investigated. The indications so far are that some crop species are much more sensitive than others, and that there are huge differences between cultivars. There is unlikely to be a problem in producing cultivars that can withstand higher amounts of UVB. In the long term, that won't be a major problem.

An increase in ozone to the concentration you mentioned would depress the yield by 10% at most for most crops in north-west Europe. However, changes in air pollution, such as those occurring in India, may be a problem. I am getting a lot of correspondence from Indian scientists who are becoming increasingly concerned. Many of their crops are produced close to the centres of population where they are actually used. More and more industry is being introduced into those areas, and if there are low chimney stacks, the dispersal of pollution emissions is poor. This needs to be taken into account in considerations of changes in productivity. Monitoring pollution in enough locations on a national scale is expensive. I believe a good ozone monitor costs $15 000.

Hoffmann: You can buy UV monitors for several thousand dollars, but higher quality, more sensitive instruments would probably cost about that.

Klein: One problem is that the relationship between increased ozone concentration and the effects on different plants is not linear. Experiments with trees suggest that there's no observable effect up to about 100 μg/m^3, but when the level is 50% higher, there are tremendous problems which might be detected not immediately but a year after. The effect will depend on circumstances, on which plant is involved, what the climate is like and how the concentration changes with time. I would be cautious about making general predictions or extrapolating from one plant to another.

Zehnder: The question is not really one of extrapolation from what we know now, but of identifying the crucial areas where knowledge is lacking. There are some areas which are more crucial for the survival of mankind and maintenance of the quality of life.

Hoffmann: One pernicious chemical which has been a problem in California is ethylene dibromide, which is used as a fungicide and has recently been banned. I mentioned earlier Bruce Ames's mutagenicity and carcinogenicity tests, in which peanut butter comes high on the scales. Highest on the Ames list of toxic chemical compounds is ethylene dibromide, in terms of both known incidences of cancer and *Salmonella* reversions in the mutagenicity tests. That's a clear case of a high impact chemical with adverse health consequences, particularly to farm workers.

Rabbinge: Ethylene dibromide is used in The Netherlands for fumigation in starch potato and bulb cultivation, but we are trying to ban it. Within the next five years it will be completely banned. I think that's also the policy of the European Community.

Sutherst: I would like to return to some of the general issues that we discussed earlier. I have been surprised at the talk of increasing temperatures increasing grain production. My understanding is that, particularly with wheat, as temperature increases the quality of the grain and its protein content decrease. This is certainly true of pastures, so a pasture-fed animal would have a lower quality of pasture at a higher temperature, which will be of significance in tropical areas.

It's not easy to separate out the effects of temperature or CO_2 on a pest without looking at what's happening to the crop the pest is on. For example, here in The Netherlands, with an increase in temperature there will probably be fewer aphid problems and fewer virus diseases in plants. However, the crops that these diseases and pests attack may also be displaced or stressed by the temperature increase. A temperature increase might increase the resistance of the plant to attack by some species of aphid, but decrease its resistance to other species. It's difficult to generalize.

The one issue that worries me more than any other is that of pesticide resistance. I guess that in 20 or 30 years time we are not going to have a problem with pesticide use, because we are not likely to have many pesticides. There is so much resistance now, so many problems with pesticide residues and so many regulations that the search for chemicals is slowing down. We are approaching a situation in which it will not be possible to move cattle to abattoirs, because they contain pesticide residues, but if enough time is allowed for the pesticide residue to dissipate, the parasites will re-infect the cattle. We tend to take it for granted that we are just going to get another pesticide. In Europe it may be feasible to reduce pesticide use through biological control and crop rotation. On a global scale, this will be more difficult for management reasons. Pests are undoubtedly more serious in tropical areas, because there are more generations per year. There are places in northern Australia, for example, where cotton production has already been abandoned because resistance to pesticides developed so quickly in pests which have many generations each year.

Pest problems will worsen with climate change. They will spread into higher latitudes. However, many of the impacts of climate change are, to me, insignificant in comparison with the problems of human population growth and the supply of technology for agriculture.

Rabbinge: I am not so pessimistic with regard to pesticides. There is still a need for pesticides in combination with the other methods. Many governments fund research dedicated to the development of resistance and biological control. Even the chemical industry is moving in that direction, not so much with parasites that control insect pests, but with viruses and fungi. I agree that this research has to progress rapidly to overcome the problems of resistance, especially in tropical areas.

Zehnder: Population growth is a major issue. The latest figures from the World Bank and the UN predict that by 2050 there will be almost 12 billion people, if AIDS is not included, or about 10 billion if AIDS is taken into account. How can these people be fed? The green revolution such as that which occurred after World War II in Europe and in the 1960s and 1970s in countries such as Mexico, Pakistan and India will not be sufficient. We need green revolutions numbers two and three. However, these revolutions must be achieved with the soil we have now. There will be no other soil available. In addition to population growth, temperatures will increase, so the tropical and sub-tropical areas will increase. Has anyone any idea how we can deal with this complex of population growth, increased pressure on soil fertility and the extension of tropical and sub-tropical areas?

Rabbinge: There have been many investigations of the potential for production in different parts of the world, and this can be quantified. Even when you're not too optimistic about realizing potential yields, the results suggest it is definitely possible to feed all those people. What is lacking is the external resources required for the productivity rise. There needs to be investment, the money to pay for these external resources. This is certainly the difficulty in parts of Asia and in most of Africa.

Avnimelech: This is also the problem in the marginal areas.

Rabbinge: The marginal areas increasingly are getting into a dangerous situation. In many countries people do not continue trying to increase yields in the better soils, but start to move into the marginal areas. In such areas there is some production for a short period, 10 years for example, then the people have to move on, because fertility has declined and production capacity has been used up and a profitable harvest is no longer possible.

Lake: Nitrogen as an external resource is something that we haven't mentioned, and this links partly with the CO_2 story, and partly with the issues Philip James raised. If plants are to show increased growth in response to increased carbon dioxide, they will also need nitrogen, for growth in the cytoplasmic component. Without nitrogen, the plants will not respond to increased CO_2. The growing world population has to be provided with protein as well as with energy. In British agriculture overall, only 10% of the nitrogen applied as fertilizer is ultimately ingested by humans. At the plant level, about half of it is taken up; if we ingested only plants, instead of using the plants to feed animals for us to eat, efficiency of nitrogen use would improve greatly. The external cost of providing the nitrogen per unit of food energy would decrease, as would pollution of the environment by wasted nitrogen. Has the issue of nitrogen been addressed in your estimates of food production?

Rabbinge: We have addressed the question of nitrogen. The waste of nitrogen and resultant pollution can be partly overcome by feeding the cattle in stables, at least during the night. Given the proper diet, not only grass but also other components which enable the cattle to digest the grass to obtain the nutritional value which they really need, the efficiency of nitrogen use can be increased. However, the problem remains that making meat or dairy products is not as efficient in terms of nitrogen use as pure plant production.

Lake: That scenario is in one way attractive, because I like my beef steak, but it contrasts horribly with the reality of Bangladesh, for example, where the only hope is an improvement in the techniques of food production coupled with recycling. We should make sure that the nitrogen we ingest and then excrete goes back on to the land, in ways which avoid the spread of disease. The idea of putting Aberdeen Angus cattle into sheds to produce high quality steaks and decrease pollution of the land seems somewhat unrealistic in terms of the problems of world food supply and developing countries.

Sutherst: I would like to challenge the idea that feeding an animal in a stall to fatten it is healthy for humans or the environment. To produce steak by feeding grain to cattle is a good way of getting a heavy carcass and hence a high fat content in the meat. Cattle raised on grass take longer to grow, but have a nutritional value vastly more healthy than that of beef produced by feedlot methods because they are usually slaughtered at a lower carcass weight. Feeding animals grain to produce a 'good' diet is, as Professor James has pointed out, not only wasteful but also unhealthy.

Reference

Jarrett RJ, Shipley MJ, Rose G 1982 Weight and mortality in the Whitehall study. Br Med J 285:535–537

Soil quality in relation to soil pollution

F. A. M. de Haan

Department of Soil Science and Plant Nutrition, Wageningen Agricultural University, PO Box 8005, Dreijenplein 10, 6700 EC Wageningen, The Netherlands

Abstract. In comparison with the environmental compartments air and water, soil is an extremely complicated system, because properties of the soil and chemical, physical and biological characteristics vary greatly between different soil systems. This makes the development of general rules for quantitative evaluation of soil quality impossible. The large variation between the buffering capacities of various soils for different compounds also complicates the question. In The Netherlands, a system of reference values for various compounds has been developed; these reference values reflect situations in which there is a low degree of contamination. For heavy metals, these reference values consist of sliding scales incorporating the influence of a soil's clay and organic matter content. Examples for derivation of effect-based soil quality standards are briefly described.

1993 Environmental change and human health. Wiley, Chichester (Ciba Foundation Symposium 175) p 104–123

Of the three environmental compartments—air, water and soil—it is the soil for which it is most difficult to assess quality in a quantitative way. This is mainly due to the fact that soils show a huge variety in composition, which in turn influences the behaviour of compounds in the soil and the effects they have on the chemical, physical and biological features of the system. Soils usually have a much larger buffering capacity than air and water. This property, as schematically indicated in Fig. 1, can be described as the capacity the soil has to allow the content of a compound, once present at a certain optimum level (I), to increase without effect. Because some potentially hazardous compounds are also necessary for proper soil functioning, these have a positive effect with increasing content at a low concentration. The area between contents I and II in Fig. 1 represents the buffering capacity. This buffering capacity differs from different compounds of interest, and with different soil properties and the varying conditions of a system that occur in practice. Moreover, these properties and conditions are usually the main factors controlling chemical speciation of a compound, which in turn plays a major role in determining its bioavailability and thus its effect.

Soil protection policy in The Netherlands is aimed at maintaining the amounts of potentially hazardous compounds in the soil as close as possible to point I of

effect

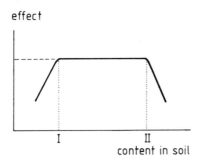

I   II
content in soil

FIG. 1.   Schematic representation of a soil's buffering capacity. The buffering capacity is the area between points I and II. A positive slope represents a positive effect, a negative slope a negative one. The range between points I and II indicates the buffering capacity. Above point II the soil is polluted. The horizontal dashed line refers to elements which are not a requirement for good soil functioning, such as cadmium, where there is no point (I) of optimal content.

Fig. 1. This will preserve part of the soil's buffering capacity for future generations, which is a condition for sustainable soil use. However, if rules and measures for soil protection are to be developed, a quantitative evaluation of soil quality is required. During the Commission of the European Communities (CEC) conference *Scientific basis for soil protection in the European Community* in 1987 (Barth & L'Hermite 1987) it was agreed that such evaluation should preferably be based on measurement of effects that can be expected from the presence and likely behaviour of pollutants/contaminants in the soil. This necessitates a quantitative risk assessment approach.

Figure 2, adapted from Van Genderen (1987), presents in a schematic way the routes by which contaminants may adversely affect humans, plants and animals, including soil organisms. In soil quality evaluation, the following question must be answered: what is the maximum quantity of a given compound in soil, C_s, acceptable if these groups of organisms are still to be protected? This question is closely related to risk analysis of health and toxicological considerations. Once the exposure–effect (more strictly dose–effect) relationships are determined, the question is transformed into one of assessment of the result of the exposure of the organism of interest.

Sometimes the exposure can be determined in a relatively simple way, as in the case of some soil organisms. The concentration (or rather the activity) in the soil moisture is usually of major importance, because the extent of exposure is the product of the time of exposure and this solution concentration. This means that the relationship between the solid phase content and the solution concentration is of prime concern. Soil chemistry provides a basis for such relationships for different contaminants and various soil systems. Adsorption and desorption isotherms are the common form of expressing them, and the

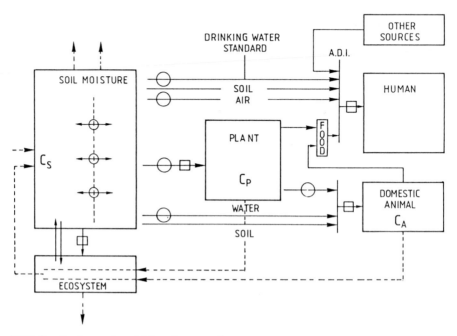

FIG. 2.   Pathways by which soil contaminants reach other compartments and groups of organisms (after Van Genderen 1987). For most compounds it is the distribution between the solid phase and the liquid phase (soil moisture) which is of prime concern. The circles represent points where information is required about transfer coefficients and the squares points where quantitative information about dose–effect relationships is required for the estimation of risks. The dotted lines indicate that a portion of the contaminant can be recycled within the soil system. C, content; A.D.I., acceptable daily intake.

distribution coefficient, $K_{s,l}$, provides a useful parameter by which to characterize interactions between the solid and liquid phases of soil.

The concentration in solution in the soil is also highly significant because of leaching into surface water and ground-water, and the uptake of compounds by plant roots. Such uptake mechanisms are generally fairly complicated (there can be, for example, preferential or discriminating uptake) so that the content of the compound in plants, C_p, cannot usually be derived simply from transpiration fluxes and solution concentration values. Thus, the liquid phase concentration is generally the most important factor, but there are a few examples where exposure, or the major part of it, at least, is directly governed by the content of the solid phase, such as where the exposure results from ingestion of contaminated soil or dust, as is the case with earthworms (Ma 1983), grazing animals (Bremner 1981), and with pica (the eating of material unfit as food) and mouthing behaviour (direct uptake of soil by children; Brunekreef 1985). Normally, however, the route of exposure is much more complicated (Fig. 2). As shown in Fig. 2, information about transfer factors of contaminants from the soil to other compartments and organisms is required, as is quantitative information about dose–effect relationships for estimation of the (risk of) effects on humans, animals, plants and the soil ecosystem. Also, a portion of the contaminant can be recycled within the system; this is true for persistent constituents such as heavy metals and slowly degrading organic compounds.

As stated in a recent report to the CEC on soil quality assessment (de Haan 1989): 'Ideally, a quantitative evaluation of soil quality requires complete information about all the transfer factors and dose–effect relationships shown [in Fig. 2]: for all compounds that can cause malfunctioning or disfunctioning of soil; for all the different soil types and soil properties that are found; and for all combinations of the different variables (e.g. pH, redox potential, accompanying compounds) that control compound behaviour in the soil system.' Because this is an impossible task at the present stage of knowledge, a number of limitations and restrictions have to be made in a first approach.

Soil quality in relation to soil function

The quality of soil is adversely influenced by contamination (pollution) of the system. The concepts of 'contamination' and 'pollution' of soil are used here in a comparable way because the difference between them is only one of the degree of damage to the soil system. Any addition to soil of compounds which may exert adverse effects on soil functioning can be defined as soil contamination. Because most soils have a certain buffering capacity, as mentioned above, it usually takes some time before the negative effects become apparent. Once they are apparent, the soil can be considered as polluted. Thus, for all practical purposes, 'polluted' means that there is malfunctioning of the soil due to a high content or availability of compounds. Such malfunctioning may refer to one specific function in particular, but sometimes also to a combination of different functions.

Some of the most important soil functions are: (i) the bearing or supportive function, for example as a children's playground or for the building of houses; (ii) the plant growth function; this can be for natural vegetation or the production of crops for animal and human consumption (with crop production, not only quantitative aspects such as yield play a role, but qualitative aspects are also of concern because consumers' health can be influenced by plant composition); (iii) the filtering function, for water, ground-water and surface water; (iv) the ecological function of soil, of which the contribution to element cycling is an important aspect. The fulfilment of these functions imposes a wide variety of quality criteria, which greatly hampers the introduction of a quality assessment methodology with general applicability and validity.

The first function, the bearing function, would probably seem the least demanding, once certain physical requirements have been met, at least. However, the experience of the 1980s and 1990s in, for example, Love Canal, USA, and Lekkerkerk, The Netherlands, has taught us that there are some minimal requirements of chemical conditions even for soil on which houses are to be built, and that use of a site for waste disposal and subsequent urban expansion is not always without problems. In the panic that arose following the discoveries of certain somewhat incautious combinations of soil usages, there was a strong demand for methods to assess the health risks for people living in such polluted areas. Of course, it was realized from the very beginning that this would require insight into the quantitative relationships between the organism's exposure to the pollutant under consideration on the one side, and the resulting effects on the other. However, at the same time it was realized that there are many different pathways of exposure (see Fig. 2). For humans, the most important routes are: soil ingestion, especially for young children; inhalation of air containing volatile polluting compounds; and drinking of water and eating of food in the form of plant and animal products. If we confine our attention to the last pathway, it becomes apparent why the evaluation of soil quality in a quantitative way is such a complicated problem. Assume that one can rely on the value of an acceptable daily intake (ADI) of a certain compound for human beings, and that there is agreement on how this ADI is allotted over different pathways of exposure; the sequence soil → plant → animal product → human, as shown in Fig. 2, means that the contribution to human exposure via animal products is akin to a fourth-order derivative of the soil quality. This complexity has undoubtedly contributed to the boom in the development of so-called multimedia exposure models that has arisen from the growing general awareness of soil pollution problems; these models have so far, unfortunately, not proved applicable in most cases for quantitative approaches.

In the stress that followed the discoveries of the, supposedly severely, hazardous urban developments built on waste disposal sites, cleaning and isolation measures were taken to restore the sites. In The Netherlands a 'Soil Clean-up' act was introduced. To provide guidelines for estimating the severity of the situation, so-called A-B-C values were suggested. These were actually intended to be indicator values with the following meaning: A-value, background value, no problems expected; B-value, indicative of a need for further investigation; C-value, indicative of severe pollution and a need for action. Although it was clearly stressed that these values were of a preliminary nature only and should be used carefully, in practice they tended to be held as rigid, scientifically based criteria which supposedly allowed a quantitative risk evaluation of the areas involved. This single-value interpretation of soil quality has generated much commotion and in many cases has led to demolition of residential areas and economic losses, the necessity of which was at least questionable.

Even though there are problems involved in a scientifically based judgement of situations like that just described, it must be realized that waste disposal sites are restricted areas of pollution. Thus, they can be considered as point sources, where control and protection against further environmental damage can be effected relatively easily. This is in contrast to large scale diffuse contamination of soil, which now is also widespread, where the possibilities for control and cleaning-up are limited. This forces the establishment of measures designed to prevent such diffuse contamination and pollution. The main sources for this type of soil degradation are invariably waste emissions into air and water, and direct addition of compounds to soil by man.

Different soil functions may come into conflict with each other, for example, when soil is used for crop production while the ground-water underneath this soil is required as a source of drinking water. With typical nitrogen fertilization of arable land on sandy soils, 200 kg nitrogen/ha/year, and a precipitation surplus of about 300 mm/year, leaching increases the nitrogen content of the ground-water to more than 120 mg/l, a concentration roughly 2.5-fold in excess of the WHO standard. In such cases, priorities have to be set with respect to soil functions or uses. Giving priority to a certain function will obviously impose limitations on other uses.

Reference values for unpolluted soil: heavy metal content as an example

During the last decade considerable effort has been exerted in The Netherlands to gain insight into the levels of heavy metals that are compatible with good soil quality. An intensive analysis was made of areas designated as nature reserves which had been managed as such for a long time (Edelman 1984). Soil of 'good' quality was then interpreted as soil that was not heavily polluted or was almost unpolluted by human activities. Obviously, a certain amount of contamination through deposition from the air is unavoidable over most of Europe, and probably all of the world. This means that it is impossible to use present samples as an indication of natural background values, for some elements, at least; the aerial deposition of cadmium and lead, for example, has been widespread.

The natural background value depends on the nature of the parent material from which the soil has developed. In The Netherlands this parent material was brought from elsewhere by wind, water or through transport with ice. During transport and deposition particles were sorted according to size, and thus at the same time according to mineralogical composition. This is of utmost importance for the occurrence of heavy metals. Sand and silt consist of primary minerals which are resistant against weathering, whereas the clay fraction (particles $< 2\,\mu$m) consists of secondary minerals formed through weathering of less resistant primary minerals. Quartz is usually the most important component of the sand and silt fraction, whereas clay minerals and hydrous oxides constitute the bulk of the clay fraction. The heavy metal content varies between the

de Haan

TABLE 1 Reference values (95 percentiles) for heavy metals, arsenic and fluoride in soil[a]

Compound	Relationship (mg/kg dry soil)	'Standard' soil ($H = 10\%$ [b]; $L = 25\%$ [c])	Ground-water ($\mu g/l$)
Cr	50 + 2L	100	1
Ni	10 + L	35	15
Cu	15 + 0.6(L + H)	36	15
Zn	50 + 1.5(2L + H)	140	150
As	15 + 0.4(L + H)	29	10
Cd	0.4 + 0.007(L + 3H)	0.8	1.5
Hg	0.2 + 0.0017(2L + H)	0.3	0.05
Pb	50 + L + H	85	15
F	175 + 13L	500	—

[a]Calculated from an analysis of topsoil from nature reserves in The Netherlands (Edelman 1984, Lexmond & Edelman 1987).
[b]H, organic matter content of the soil (% weight).
[c]L, clay content of the soil (% weight).

different minerals—the heavy metal content of quartz is extremely low—such that the heavy metal content of sediments is related to their texture (grain size distribution). The natural background contents for different sediments are fairly well known. They may be considered as the natural background because the sediments were formed and deposited before human activities influenced heavy metal distribution. Of course, the original background values have changed since that time because of processes of soil formation and later human activities, particularly the industrial revolution.

The values gathered through sampling of soil from nature reserves should therefore be termed 'current background values', to indicate that they are the original values as modified by soil-forming processes and deposition from the air. To treat these values as 'reference values for good soil quality', as is at present usually done in The Netherlands for policy reasons, is, as a matter of principle, not correct, because this would suggest that there is information about desirable metal content and soil functioning even though, as stated above, such information is still lacking. Moreover, it might well be that the levels are below optimum for specific soil functions; in crop production, for example, certain levels of those heavy metals which are trace elements for plant growth are necessary. It is therefore preferable to characterize the concentrations found in soil from nature reserves as 'present background values for not abnormally contaminated soils'.

Lexmond & Edelman (1987) have developed a model for the quantitative description of the influence of the clay content and the organic matter content of a soil on the occurrence of heavy metals, with or without human influence on metal content. This model has been used for an in-depth analysis of the data originally collected by Edelman (1984) from topsoil (0–10 cm) from nature reserve areas (Lexmond & Edelman 1987). These new reference values for heavy metals, arsenic and fluoride and their relationship with the organic matter (H) and clay (L) content of soil are shown in Table 1. The formulae used to calculate the reference values as presented in Table 1 have been derived from the inventory data mentioned above (Edelman 1984). As an example, the values for a 'standard' soil with 25% clay and 10% humus are also shown.

The Institute of Soil Fertility at Groningen, The Netherlands, has made a comprehensive inventory of the heavy metal content of Dutch arable soils (Van Driel & Smilde 1982, Wiersma et al 1985). These values for topsoil samples (0–20 cm), again as a function of the clay content of the soil samples, can be compared with the data from the nature reserve areas. This comparison is made in Fig. 3, for the metals Cr, Ni and Cd. As shown in Fig. 3, for Cr and Ni the situation in arable soils is fairly comparable with that in nature reserve soils. For Cd, in a number of cases the arable soils show higher values, which is probably attributable in part to the use of phosphate fertilizers and in part to the fact that large areas of The Netherlands have been contaminated by zinc smelter emissions. The same comparison for the metals Cu, Hg and Zn would reveal many more situations where the amounts in arable soils are higher than in the nature reserves.

It cannot be emphasized enough that this approach, although a valuable means of providing a preliminary rough evaluation of the degree of soil contamination, is by no means an effect-based or functionally oriented means of soil quality assessment. Such assessment should consider not only the mere presence of a compound but also its behaviour, especially with respect to bioavailability.

Examples of effect-based soil quality standards

When developing standards for soil quality in relation to soil functions, it might be reasoned that protection of the most susceptible function would automatically safeguard all other functions. However, our current knowledge is insufficient even to allow us to differentiate between degrees of susceptibility. It therefore seems a more realistic approach to consider a specific soil function as a first priority in a certain case, and then to try and develop soil quality standards (conditions, requirements) needed to permit this function. Even with this limitation, the complexity of the system is such that there are still problems to be solved.

A few examples of functional standards for soil quality have been described. Lexmond (1980, 1981) derived a toxicity index for copper in soil with respect to plant growth. The availability of copper for uptake by plant roots appears to be controlled by three main factors, the copper content of the soil (measured in g/kg soil after extraction with 0.43 M HNO_3), the organic matter content

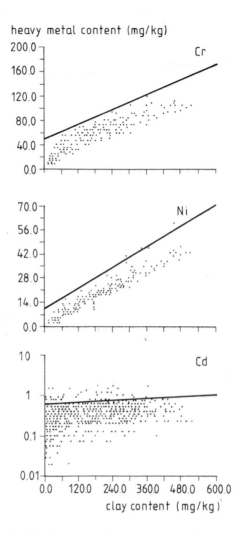

FIG. 3.   The Cr, Ni and Cd content of Dutch arable topsoils (0–20 cm) in relation to the clay content. The lines plot the upper extent of the concentrations of these heavy metals in topsoil (0–10 cm) from Dutch nature reserves (see Table 1).

of the soil (g organic carbon/kg soil) and the pH (as measured in a suspension of $CaCl_2$ solution). The latter two factors are of direct importance to copper binding in soil: as organic matter content and pH increase, copper ions in solution in the soil are increasingly adsorbed onto the solid phase of the soil and are less available for uptake by plant roots. The toxicity index is shown in Fig. 4, together with some field data.

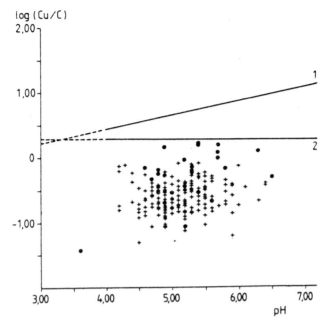

FIG. 4.   An illustration of the application of Lexmond's (1980, 1981) index for copper toxicity with respect to plant (in this case, maize) growth. The points are field data collected from samples of sandy soils. Line 1 represents the toxicity index. At values above these lines the relationship between copper content, organic matter content and pH of the soil is such that soil fertility is reduced enough to cause a yield depression of 10% or more. Line 2 shows the relationship in a soil with a Cu : C ratio of 2, where damage to soil fertility is unlikely over the range of pH values of interest in agriculture (pH 4.5–7.5).

Contents of Cd in soil at which there is acceptable leaching to ground-water and surface water have been derived (de Haan et al 1987). The Freundlich equation satisfactorily describes cadmium adsorption in the practically relevant concentration range. This equation is shown in (1), where q is the amount adsorbed onto the surface of soil particles, c the concentration in solution, and k is an empirical parameter, the so-called adsorption coefficient; the value of n appears to be fairly constant, approximately 0.8, under all conditions of interest.

$$q = kc^n \qquad\qquad (1)$$

In laboratory adsorption experiments, the values k and n were assessed for several different soils. In spite of the considerable problems in experimental accuracy, a number of system parameters that were not sufficiently included in soil type differences were varied. These parameters were ionic strength (I_O),

de Haan

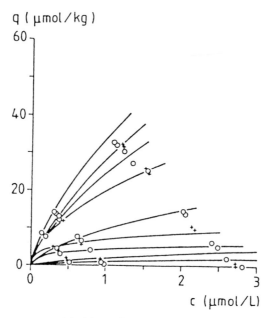

FIG. 5.   Cadmium concentration (q) as a function of the concentration of Cd in solution (c) at different values of ionic strength (I_O), [Cl^-] and [Ca^{2+}], to illustrate the point that all relevant factors need to be taken into account before arriving at the data that are to be used in environmental protection.

solid:solution ratio, electrolyte composition (including the competing heavy metals zinc, lead and copper), pH, etc. As an example of the large differences encountered for one specific soil at different [Cl^-], I_O and [Ca^{2+}], the experimental and fitted isotherms are shown in Fig. 5. This clearly demonstrates the huge influence these three parameters have on cadmium adsorption. Care is thus required in assessing the adsorption parameters for a particular case because realistic values of the system parameters may differ between, for example, agricultural soil water and waste percolation water. The major factors controlling cadmium adsorption were taken into account in the derivation from (1) of an empirical relationship (2), in which $a \approx -0.48$ and $b \approx -0.41$ held for a wide range of soil types. For the effects of ionic strength (I_O) and

$$q = k_1 [H^+]^a [Ca^{2+}]^b c^n \tag{2}$$

cadmium chloride complex formation to be taken into account, the concentration term (c) in (2) should be replaced by the Cd^{2+} activity (Chardon 1984). Even for one soil type the value of k_1 varied significantly under different conditions. The equation shown in (3) was derived to describe the relationship between adsorption and the predominant adsorbing phase (organic carbon, oc) and

$$q = k^* \text{oc} \, [\text{H}^+]^a c^n \qquad (3)$$

pH; oc is expressed as % (w/w) ($100 \times$ mass of organic carbon/mass of dry soil) and $[\text{H}^+]$ in M. In (3), $a \approx -0.48$ and oc varies from 0.8 to 4.0% (w/w). The value found for k^* is on average 2.37×10^{-5} (with q expressed in mol kg^{-1} and c in mol m^{-3}). From (2) and (3) one can see that as system parameters may vary in space, this is also the case for concentrations in equilibrium with a designated total cadmium content (ρT) as given in (4).

$$\rho T = \rho q + \theta c \qquad (4)$$

The definition of acceptable total contents in soil of contaminants should preferably be based on the relationship between total content (T) and soil quality. Thus, with (4) in combination with (1) the acceptable total cadmium content in soil may be calculated for various commonly accepted standards for cadmium concentrations in ground-water and surface water. As an illustration, the following standards are considered (in the units commonly used in this context): (i) the Dutch standard for surface water to be used for preparation of drinking water, 1.5 μg/l; (ii) the Dutch advisory value for surface water, 2.5 μg/l; (iii) the EC standard for drinking water, 5.0 μg/l; (iv) the FAO and WHO standards for drinking water, 10.0 μg/l. The total contents in equilibrium with these concentrations are given for different values of the adsorption coefficient, k, in Table 2, which shows the resulting pronounced variation in acceptable total cadmium contents. The contents given in Table 2 can be evaluated with reference to the common total cadmium content of Dutch agricultural topsoils, $T < 0.3$ mg kg^{-1}. A soil with a total content exceeding 1 mg kg^{-1} is considered to be heavily contaminated.

In both these examples, the copper toxicity index and the cadmium standards, a number of years of intensive research work was required to collect the data. It became clear that, as expected, a single value approach to the evaluation of soil quality is useless, and that sliding scales are required because of the variation in the properties and conditions of soil systems.

Some remarks on bioavailability of compounds in soils

The concept of bioavailability is important with respect to soil quality because it links, in principle, the behaviour of the compound and its effect on organisms. At present, these links are not well established and a precise scientific definition of the word is not simple to derive. Aspects of bioavailability and soil quality have been discussed by Van Riemsdijk (1990); some of these aspects are discussed below.

Bioavailability is the availability of a chemical with respect to a specific organism, and can be measured only in terms of the effect the chemical has on an

de Haan

TABLE 2 **Acceptable cadmium contents of soil for different standards for cadmium concentration in solution in the soil at varying values of $k(\mu g^{1-n}\ l^{n}\ g^{-1})$ and $n = 0.8$**

k	Acceptable Cd content of soil (mg/kg) for various Cd water standards			
	$1.5\ \mu g/l$[a]	$2.5\ \mu g/l$[b]	$5.0\ \mu g/l$[c]	$10.5\ \mu g/l$[d]
0.001	0.001	0.002	0.004	0.006
0.01	0.01	0.02	0.04	0.06
0.1	0.1	0.2	0.4	0.4
0.3	0.4	0.6	1.1	1.9
0.5	0.7	1.0	1.8	3.2
0.9	1.2	1.9	3.3	5.7
6.0	8.3	12.5	21.7	37.9

[a]The Dutch standard for surface water to be used for preparation of drinking water.
[b]The Dutch advisory value for surface water.
[c]The EC standard for drinking water.
[d]The FAO and WHO standards for drinking water.

organism. This effect is in general a complicated function of the concentration of the specific chemical as present in a certain complex matrix. The matrix can be, for example, a particular soil or sediment. The effects that can be measured include growth, uptake, respiration, degradation, root development or enzymatic activity. It has been suggested that when the chemical is of inorganic nature, bioavailability should be defined in terms of the concentration of the inorganic element (Cu, N, P, Al, etc.), whereas for organic pollutants a specific chemical species (e.g. HCH) should be used. Bioavailability can thus be defined as the availability of an element or organic species as present in a (complex) matrix in relation to the likelihood of it causing an effect (positive or negative) in a specific organism. The bioavailability of an element present in a specific form depends on the type of organism and on the form of the element. For blue algae and leguminous plants N_2 gas may be fully available whereas for most plants the bioavailability of nitrogen in the gaseous form is zero, although nitrogen in the form of nitrate may be fully bioavailable. Thus, the bioavailability of an element within different chemical species can be completely different. The bioavailability is also dependent on the composition of the matrix in which the chemical is embedded. There can be, for example, competitive effects with other species present in different amounts in various soils, differences in pH, reactive surfaces, etc. The formation of soluble complexes between a metal and organic ligands such as humic and fulvic acids can increase or decrease the bioavailability of the metal; the bioavailability can increase even when the metal–organic complex itself is non-bioavailable if transport of the metal to the organism is limiting the bioavailability, because complexation with dissolved ligands leads to an increased total metal concentration in solution, facilitating transport to the surface of the organism, where the metal may exert its effect after dissociation from the complex. The organic–metal complex in this case is acting merely as a carrier. A decrease in bioavailability may result from complex formation when the main effect of the complexation is a decrease in the (available) free metal concentration in solution.

The pH may also greatly affect the bioavailability of an element, and this effect can depend on the composition of the matrix. In a culture solution, the availability or toxicity of copper, as measured by reduced growth, increases with increasing pH, that is, with decreasing proton concentration (Lexmond 1980), because decreased competition between protons and metal ions at higher pH for surface sites on the organism (root, cell wall) increases the surface's affinity for copper. However, for plants in soils, copper toxicity or availability generally decreases with increasing pH. In the plant–soil system, the copper partitions over the solution phase, the surface of the biota and the surface of soil colloids. With an increase in pH the preference of copper to bind to the surfaces (both living and dead) will increase because of decreased competition with protons. If the preference for the soil surfaces is increased more than that for the surface of the biota, a decrease in bioavailability will result from an increase in pH.

Another interesting and complicating aspect of bioavailability is that organisms may actively influence bioavailability, positively or negatively, by changing the chemical composition of their immediate environment. For example, the excretion of organic ligands by earthworms decreases the bioavailability of lead and cadmium, as assessed by measurement of their uptake by the worms from aqueous media, because these organic ligands bind the metals effectively (Kiewiet 1989). This illustrates that a rather detailed knowledge of chemical interaction processes is required for interpretation of the effect of environmental factors on bioavailability. What is needed are generalizable concepts of bioavailability that can be applied to any soil, sediment or aqueous system so that the effect of environmental factors on the bioavailability can be interpreted, predicted and modelled. The examples given indicate that reaching this goal is not a trivial problem.

References

Barth H, L'Hermite P (eds) 1987 Scientific basis for soil protection in the European Community. Elsevier Science Publishers, Essex, UK

Bremner I 1981 Effects of the disposal of copper-rich slurry on the health of grazing animals. In: L'Hermite P, Dehandschutter J (eds) Copper in animal wastes and sewage sludge. D Reidel, Dordrecht, p 245–255

Brunekreef B 1985 The relationship between environmental lead and blood lead in children: a study in environmental epidemiology. PhD Thesis, Agricultural University, Wageningen

Chardon W 1984 Mobiliteit van cadmium in de bodem (Mobility of cadmium in soil; in Dutch). Staatsuitgeverij, The Hague (Soil Prot Ser 36)

de Haan FAM 1989 Research priorities for soil quality assessment. In: de Haan FAM
    et al Soil quality assessment: state of the art report on soil quality. EC, Directoral
    General XII, Brussels, p 1–18
de Haan FAM, van der Zee SEATM, van Riemsdijk WH 1987 The role of soil chemistry
    and soil physics in protecting soil quality: variability of sorption and transport of
    cadmium as an example. Neth J Agric Sci 35:347–359
Edelman Th 1984 Achtergrondgehalten van stoffen in de bodem. Staatsuitgeverij, The
    Hague (Soil Prot Ser 34)
Kiewiet A 1989 Accumulation of lead and cadmium by *Lumbricus rubellus*. Internal
    report, Research Institute for Nature Management, Arnhem
Lexmond ThM 1980 The effect of soil pH on copper toxicity to forage maize grown
    under field conditions. Neth J Agric Sci 28:164–183
Lexmond ThM 1981 A contribution to the establishment of safe copper levels in soil.
    In: L'Hermite P, Dehandschutter J (eds) Copper in animal wastes and sewage sludge.
    D Reidel, Dordrecht, p 162–183
Lexmond ThM, Edelman Th 1987 Huidige achtergrondwaarden van het gehalte aan een
    aantal zware metalen en arseen in grond. In: Handboek voor milieubeheer, deel bodem.
    Samsom, Alphen aan den Rijn (Hoofdstuk D4110) p 1–35
Ma W-c 1983 Biomonitoring of soil pollution: ecotoxicological studies of the effects
    of soilborn heavy metals on Lumbrist earthworms. In: 1982 Annual report. Research
    Institute for Nature Management, Arnhem, p 83–97
Van Driel W, Smilde KW 1982 Heavy metal contents of Dutch arable soils. Landwirtsch
    Forsch Sonderh 38:305–313
Van Genderen H 1987 Relatie tussen bodemkwaliteit en effecten. Verslag Studiedag
    Bodemkwaliteit. Technische Commissie Bodembescherming, The Hague
Van Riemsdijk WH 1990 Introduction on soil chemical aspects of bioavailability. Caput
    series soil chemical aspects of bioavailability. Department of Soil Science and Plant
    Nutrition, Wageningen Agricultural University, Wageningen
Wiersma D, van Goor BJ, van der Veen VG 1985 Inventarisatie van cadmium, lood,
    kwik en arseen in Nederlandse gewassen en bijbehorende gronden. Instituut voor
    Bodemvruchtbaarheid, Haren (Rapp 8-85)

DISCUSSION

Kleinjans: You mentioned the direct ingestion of soil by children. In quantitative risk assessment, this intake of the soil, particularly by children, represents the main calculated intake of a pollutant. Do you know whether the heavy metals, for example, which are naturally present in soil but also may be present as pollution, are biologically available in these children?

de Haan: They are bioavailable—that's one of the problems. There is a shunt between the soil and the human body. When we measure the heavy metal content of soil in the laboratory, the first thing we do is extract the soil sample with acid, which is what children and adults and animals do when they take up the soil particles. This is an efficient way of extracting heavy metals from the soil particles, after which they can be taken up directly into the bloodstream. Van Wijnen has done good work in this area. We now know that children playing outdoors take in on average around 200 mg soil/day. If you know something about the heavy metal content of a specific soil, you can start to calculate the intake of heavy metals. Usually, this tells you that there's not much risk, at least not in the residential areas. A more dangerous situation is where people are growing three or four crops in their allotment gardens and do not control the pH of the soil. In The Netherlands these allotment gardens are situated in areas where there is traffic, along railways or roads. That is where problems with lead and cadmium arise.

Klein: Direct uptake of heavy metals from soil is known to be a problem, but uptake of dioxins or other hydrophobic compounds is less well understood. Last summer, questions were raised by the media in Germany about the safety of playground material prepared from residues from industrial plants. The attitude of the government of North Rhine Westphalia was very fair. They decided to assess the concentrations of dioxins in the blood of children and adults. Secondly, they asked an independent body of toxicologists to set an acceptable concentration limit for blood dioxin before the data from these people were available. The analysis showed that dioxin levels in the children from these areas were well below the maximum level set by the toxicologists earlier. The problem was resolved in a rational way, without a lot of money being spent to clean up these areas unnecessarily.

As you stressed, the nature of the compound must be considered, and these problems can be dealt with if they are approached in a rational manner. You mentioned three aspects that must be considered when trying to represent toxicity with one value. Such a value is specific to each compound, to the nature of the soil, and to parameters such as pH. In any assessment you should include a fourth variable, the use of the land—whether it is a playground, an industrial site, a residential area or a drinking water supply area.

de Haan: I fully agree with you; that's what I meant by soil function.

Edwards: How well do earthworms tolerate polluted soil? Can they be used as an indicator of soil quality?

de Haan: They are indicators, but they are not good indicators. Arthropods, for example, are much better. Earthworms are sensitive to copper. If you want an idea about the general condition or quality of the soil, arthropods are better organisms to study.

Vogel: You showed that we shouldn't use up the buffering capacity of the soil, that we should remain close to point I on Fig. 1. What would you recommend should be done if you are coming from the other side, where the soil is already polluted beyond point I? Do you think, bearing in mind the economic factors, that the soil should be cleaned up?

de Haan: This is really a political issue. I think personally that we should spend our money in a better way than that which is being promoted at the moment. The requirement in The Netherlands is that the soil is to be brought back to the A value, the reference value. This is, in most cases, impossible, and where it is possible, the cost of removal of the last 10–15% is about 95%

of the total cost. There is little sense in this attitude, because after such remedial action, you are left with no real soil anyway! What's left is not soil because the organic matter has been burnt out and the structure of the soil has been destroyed. If it is to be used in building material, for example, it would be better to put a high requirement on the leaching tendency.

Vogel: For remediation projects we may define a new reference point.

de Haan: In The Netherlands you will get government money only if you bring the soil back to the reference value.

Zehnder: This is because all soil has to fulfil the multifunctionality requirement. Such a strict requirement prevents the cleaning up of many polluted soils. In The Netherlands, a soil under a road has to be of the same quality as that in a children's playground or in domestic gardens. The standards have recently begun to be relaxed and specific standards are now being applied, according to the future use of the polluted soil.

Avnimelech: This approach to soil pollution differs from that of the Environmental Protection Agency (EPA) in the USA. The European view is, as Professor de Haan explained, that the buffering capacity of the soil should not be used up and the amount of a contaminant added to the soil should not exceed the amount that can be taken out of the soil by plants.

de Haan: And I do realize that this is impossible, which is why we need to maintain some buffering capacity.

Avnimelech: The EPA's approach, taking it to the extreme, is not to care about the quality of the soil as long as the consumer, human or plant, is not endangered, because there are processes which will prevent the pollutants getting into the consumer, processes of absorption, precipitation, mobility, time, whatever. If during lunch I throw my cutlery out of the window I will have polluted a whole acre, according to the European approach, because the amount of nickel, zinc and cadmium in the cutlery would exceed the limits. However, it is not accessible to the consumer. Perhaps the EPA's approach is more realistic in some circumstances.

Any mechanistic approach centred around a given number, one number, makes us blind to the effects of other factors, whether acidification or global climatic change or whatever, on soil contamination.

Another important point is that the values mentioned by Professor de Haan are relevant to climatic conditions and condition of the soils in The Netherlands. In semi-arid regions, where leaching is much less, the natural background values are way above those found in The Netherlands, above the B level or even the C level. However, because of the high pH and the differing quantities of $CaCO_3$, the bioavailability is low. We have to realize that such a mechanistic approach cannot be applied over a wide range of conditions.

Zehnder: As we discussed earlier, the salmon is the indicator organism for good quality water in the River Rhine. Whether the salmon can be eaten is not yet an issue. In the great lakes in the USA there are salmon but it's not advisable to eat them because they contain high concentrations of polychlorinated biphenyls. For soil, there is no indicator organism or parameter. What is needed is a simple parameter by which to describe a certain soil quality. As you said, what really matters is not the concentration in the soil, but what finally gets into the plant and into ground-water, and also how the soil ecosystem is affected.

de Haan: There are very few such soil quality standards. As I described, we have developed one in The Netherlands for the phytotoxicity of copper. In this, we take into account the three factors that are relevant: the copper content of the soil, the organic metal content of the soil, and the pH of the soil. The other relevant factor is the way in which the soil is used. We are beginning to take the most important factors into account, but there is a long way to go. The derivation of this phytotoxicity standard for copper in relation to maize growth took five years of scientific work.

James: I don't understand about the techniques that are available for reclaiming soils. Earlier (p 60), you talked about huge volumes of soil being dumped in some parts of The Netherlands. Now you are talking about the extraordinary cost of reclaiming soil and reducing the content of heavy metals. How is this done?

de Haan: In The Netherlands there are soils which are considered to be too highly polluted by heavy metals, such as the dredged soils I referred to earlier (p 60). Because we do not have the money or the techniques to reclaim soil on this scale, these millions of tonnes of dredged soil, we are now considering possible ways to store this material safely. That's what I was discussing there. At the same time, there are locations where the levels of cadmium or lead, for example, are too high, where the soil has to be cleaned up. Here, we extract heavy metals with artificial organic complexing chemicals such as EDTA and TPA.

James: You are not then manipulating the bioavailability, dealing with those other parameters you mentioned; you are actually trying to physically extract the pollutant so that its concentration goes down.

Edwards: An efficient chelating compound would solve the biological problem. The only approach I can envisage with a vast field is to add chelating substances which are themselves harmless and generate stable compounds, and leave them in. This would solve the problem at the biological level, and then it would be a matter of getting the legislation to conform to the biology.

de Haan: I agree with you. There is no sense in restoring the soil to this 'reference value'. We have extracted heavy metals with EDTA and TPA. It's relatively easy to extract around 45% of the heavy metals. However, even after repeated extraction there is around 50% left, but if this is apparently not available, why should we worry about it? If you can't extract it with complexing agents or with acids it must be very strongly bound and is thus safe in that system. Of course, you have to be sure that the bioavailability has been reduced to an acceptable level.

Mansfield: Is there enough communication between different groups of policy-makers with regard to the management of soils? Take the example of acidification. Acidification is known to be worst in forest regions, where the rough aerodynamics of the forest canopy cause a greater deposition of acidic pollutants from the atmosphere. In forest regions, acidification is worst where the buffering capacity of the soil is low. That has been described by the forest science community. But another community of people are looking for alternative uses for agricultural land, and are proposing that we should plant more trees, without regard for what the consequences might be. Are there overall strategies being debated and applied? Are people even thinking about the consequences of some of these actions, which might be rather widespread in Europe over the next decade or two?

de Haan: There is a real lack of communication, even at the scientific level, between different disciplines. Another problem is that of compatibility of standards. There is no sense in bringing in soil protection legislation when you are unable to take care of air quality. There is a lack of scientific communication, but a much greater lack of communication between policy-makers in different countries, and also between the policy-makers and the scientists.

Klein: There will shortly be a meeting in Germany at which the representatives of at least five different ministries on the federal level will meet with the members of scientific institutions reporting to these ministries to discuss what is being done in soil research and what needs to be done. The ministries of science and technology, of the environment, of commerce, of agriculture and of health will be represented. In Germany, the political and administrative view is very much in favour of the Dutch A-B-C reference system. If you are rejecting this system on scientific grounds, just as Germany is about to implement it, we have a problem. In Germany and in Europe, soil legislation is moving rapidly, and there will be steps towards a new law on soil protection in Germany within the next year or two.

Hoffmann: Are there any problems in Europe with selenium? Extraction of selenium from the subsoil, because of irrigation practices, has become a major problem in California. The selenium is extracted from materials far below the soil level in draining irrigation water which is then brought back to the surface, highly loaded with selenium IV (selinite), which is toxic to birds and aquatic organisms (Bowie & Grieb 1991).

de Haan: This isn't a problem in The Netherlands, as far as I know, but problems with selenium are only now being reported in the literature.

Elliott: Children eating contaminated soil is not the only health concern. In the UK there is much public concern about waste dumps. The mechanism by which adults might be affected is much less clear than with air pollution. In southern California there have been large studies on the association between toxic waste dumps and cancers, with essentially negative results. We need to consider adult health as well as child health.

de Haan: We really should make quantitative risk assessments in such cases. We are doing this now where there are residential areas on waste dumps. However, this is where the A-B-C system hampers us. In the past, when levels of trichloroethane in houses exceeded the C value, the houses were demolished. No one appreciated that a dry-cleaned suit will cause great exposure to trichloroethane. The problem is one of quantitative risk assessment.

Elliott: That's true, but public perception of risk, concern about the soil, is also important. The task is not only the scientific part, but also involves dealing with public concerns. Determining whether or not cancers are more common on a site is an important part of the response.

Zehnder: Risk assessment is important, but I don't think houses *should* be built on a dump site, whether or not there is a risk we can forsee. Living on clean soil in the mountains is still better than living on the site of a dump!

Reference

Bowie GL, Grieb TM 1991 A model framework for assessing the effects of selenium on aquatic ecosystems. Water Air Soil Pollut 57:13–22

Arthropods as disease vectors in a changing environment

R. W. Sutherst

CSIRO Division of Entomology, Cooperative Research Centre for Tropical Pest Management, Gehrmann Laboratories, University of Queensland, Brisbane, Queensland 4072, Australia

Abstract. Arthropod vectors need to acquire energy, moisture, hosts and shelter from their environment. Changing human populations and industrialization affect almost every aspect of the environment. In particular, the prospects of climatic warming, urbanization and vegetation changes have the potential to materially affect global patterns of vector-borne diseases. Global warming will enable the expansion of the geographical distributions of vectors. The population dynamics of vectors will change in response to extended seasons suitable for development followed by less severe winters. The incidence of epidemics is likely to change in response to an expected disproportionate increase in the frequency of extreme climatic events. The impact of such changes on each of the major vector-borne diseases is reviewed and projections are made on the likely global areas at risk from spread of disease vectors. Research needs are identified and response strategies are suggested in the context of the ever-increasing impact of human populations and industrial activity on the environment.

1993 Environmental change and human health. Wiley, Chichester (Ciba Foundation Symposium 175) p 124–145

There are a large number of arthropod vectors of human diseases, the most important of which are listed in Table 1. Mosquitoes are vectors of many nematode, protozoan, rickettsial and viral diseases of humans. They breed in open water in temperate and tropical zones. Ticks are widespread acarines which transmit many viral, rickettsial and protozoan pathogens to humans. Tsetse flies are African biting flies which occupy much of the continent's savannah country and act as vectors of trypanosomes which infect both livestock and humans. Triatomine bugs are best known for their vectorial role in Chagas' disease (American trypanosomiasis) in the Americas where some species are well adapted to mudbrick and straw dwellings. In addition to the above groups there are many species of biting flies which transmit a range of pathogens to humans. Here, black-flies (*Simulium*) and sand-flies (*Phlebotomus*) will be considered only briefly along with fleas and body lice; these insects are not considered to be sensitive to environmental change.

TABLE 1   Arthropod vectors of human diseases

Vector	Disease	Pathogen
Mosquitoes (*Anopheles* spp.)	Malaria	*Plasmodium* spp.
Mosquitoes (*Aedes* spp.)	Dengue fever Yellow fever	Virus Virus
Mosquitoes (*Aedes* and *Culex* spp.)	Encephalitides Rift Valley fever	Viruses Virus
Sand-flies (*Phlebotomus*)	Leishmaniasis (Kala-azar etc.)	*Leishmania* spp.
Triatomine bugs	Chagas' disease (American trypanosomiasis)	*Trypanosoma cruzi*
Tsetse flies (*Glossina*)	African trypanosomiasis	*Trypanosoma* spp.
Black-flies (*Simulium*)	Onchocerciasis	*Onchocerca*
Mosquitoes (*Culex, Anopheles, Mansonia and Aedes* spp.)	Lymphatic filariasis	*Wuchereria* spp., *Brugia* spp.
Ticks	Colorado tick fever Tick-borne encephalitis Tick typhus Lyme disease	Virus Virus *Rickettsia* spp. *Borrelia burgdorferi*

The environment of a vector can change in many ways, but the major changes relate to the availability of energy, water, shelter and hosts. These requirements may be affected by either systemic global changes such as climate change or a series of localized but widespread cumulative effects such as water storage, drainage or irrigation, vegetation clearing or regrowth, and changes in human and animal population densities and movements. The potential impact of such changes on each major group of vector is reviewed.

The impact of climate change on mammalian or avian intermediate hosts of vector-borne diseases is likely to be small compared with that on the arthropod vectors themselves. Nevertheless, the behaviour of water birds and the population density of some rodent hosts are greatly affected by climate. These changes can have significant impacts on the incidence of diseases such as Murray Valley encephalitis in Australia and plague in southern Africa.

The extent and timing of the mooted increase in the greenhouse effect are very uncertain. One scenario is used in this exploration of the potential nature of the impacts of climate change. Temperature is assumed to increase by 1 °C per 10° latitude (for example, by 5 °C at a latitude of 50 °N) with a doubling of the CO_2 concentration expected around the year 2030. Further, a global increase of 20% in summer rainfall and a 10% decrease in winter rainfall is assumed (Pittock & Nix 1986). A computerized system, CLIMEX (Sutherst & Maywald 1985, Sutherst 1991), is used to incorporate the temperature and moisture requirements of a species. This information is then used to investigate the impacts of this climate change scenario on vector populations and geographical distributions. CLIMEX describes the response of an animal or plant species to climate, without relying on detailed biological data. It produces a growth index (GI) to describe the weekly and annual potential for population growth and an ecoclimatic index (EI) which integrates the GI with the limiting effects of extreme climatic conditions.

The impacts of environmental change

Climate change

There is a wide range of responses of different vector groups and diseases to climate. At one extreme are the triatomine bugs and fleas, which are insensitive to climate because they have adapted to living in human dwellings which provide a constant, favourable environment. Similarly, holes in termite mounds and rodent burrows provide a buffered microclimate for sand-flies, and fast-flowing streams do the same for black-flies. Of the ectoparasites, body lice are the most insulated from the external environment.

Hoogstral (1981) reviewed the changes taking place in patterns of tick-borne diseases in humans in response to environmental changes. Many viruses have changing patterns of transmission. Some tick-borne pathogens are already widely distributed and are not likely to undergo a major shift in distribution as a result of projected global warming. For example, *Rickettsia rickettsii*, the causative agent of Rocky Mountain spotted fever in North America , is transmitted by two species of *Dermacentor* which together cover most of the continent (McDade & Newhouse 1986). A similar situation applies to Lyme disease, which is caused by a spirochaete, *Borrelia burgdorferi*, that is transmitted by different species of ticks around the world. Foci of heavy infestations are associated more with the availability of alternative hosts and vegetation than with climate. Recent increases in the incidence of Lyme disease have resulted from the swing towards conservation of natural fauna and flora.

Insensitivity to the predicted changes in climate is also likely to be shown by some species of tsetse fly which transmit trypanosomes in tropical Africa. This conclusion is based partly on the assumption that the extent of climate change in tropical areas will be less than that in temperate regions. Tsetse are likely to undergo only a small shift in distribution, except in Southern Africa where the opportunity exists for invasion of eastern coastal areas if the climate warms significantly. Migration into highland areas in response to climatic warming is possible, but will depend on the amount of vegetation remaining in these already highly populated areas. The west African species of tsetse fly *Glossina palpalis* is limited in the northern part of its range to areas with monthly average maximum temperatures below 36 °C (Fig. 1a). The expected temperature changes, combined with predicted increased summer rainfall, will have little effect on its distribution (Fig. 1b). The increased rainfall could lead to an increase in vegetation cover and a consequent spread of the flies northwards beyond the limit expected to be set by temperature alone.

At the other extreme are mosquitoes, which are dependent on surface water in which to breed, and pathogens such as *Plasmodium* spp., which have minimum temperature requirements for sporogony—*P. vivax* and *P. falciparum* need a temperature above 16 °C and 20 °C, respectively (Weihe 1991)—and transmission is most rapid between 25 and 30 °C. The current global distribution of malaria transmission throughout the year or seasonally has been described by Dutta & Dutt (1978). The actual distribution is more restricted than it could be because of successful eradication efforts in areas such as Europe and the failure of some vector species to colonize all areas that are climatically suitable to them. On the basis of this distribution and historical records in Europe, a global distribution of regions in which conditions are suitable for transmission of malaria can be established; this is shown in Fig. 2a. As shown in Fig. 2b, the projected global warming would substantially increase the area at higher latitudes suitable for transmission (see Fig. 3 for detailed maps of Europe). The geographical distribution of malaria is probably determined largely by the temperature needs of *Plasmodium* for sporogony (WHO 1990), with additional limitations to transmission in some regions being set by the poor vectorial capacity of the local mosquito fauna, abundance of livestock which act as alternative hosts and so break the cycle of transmission of disease between humans, or drainage of marshlands (de Zulueta et al 1975). There are about 50 species of mosquito which can act as vectors of malaria, so it is not feasible to infer climatic requirements for permanent persistence of this disease on a global scale with any degree of confidence without a detailed analysis of the requirements of each vector species, which is beyond the scope of this paper.

There has been a temperature rise of 0.5 °C over 25 years at Tanga in the coastal lowlands of Tanzania and over the same period the infection probability of mosquitoes has more than doubled. Lines et al (1991) found no relationship between the infection probability and seasonal temperatures, thus excluding the temperature change as a likely explanation for the increase; the rise was instead attributed to changes in drug use. Given that temperatures in Tanga are usually well above the threshold for development of both the vector and the parasite, the lack of an effect of temperature is not surprising. Increases of temperature of this order under global warming are therefore unlikely to affect the epidemiology of malaria greatly in areas where it is currently endemic, but such warming is likely to affect areas adjacent to endemic areas that are currently protected by insufficient heat accumulation.

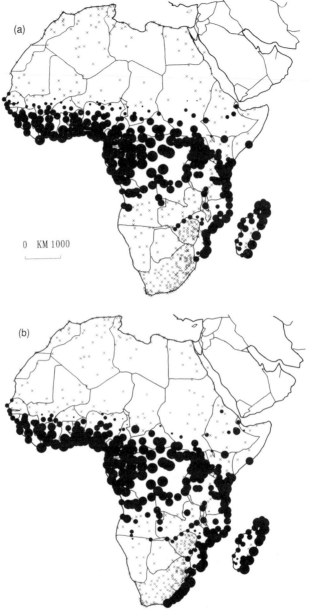

FIG. 1.   The geographical distribution of climates suitable for permanent persistence of *Glossina palpalis*, a tsetse fly which transmits trypanosomes, estimated using CLIMEX, (a) based on present climate and (b) under the climate change scenario described in the text (a 1 °C increase in temperature per 10° latitude coupled with a 20% increase in summer rainfall and a 10% decrease in winter rainfall). The size of the circles is proportional to the favourableness of the climate (CLIMEX ecoclimatic index, EI) for persistence of the tsetse fly. Crosses indicate places where the climate is not suitable for the fly to live (EI = 0).

The global distribution of *Aedes albopictus*, a freshwater-breeding mosquito and recognized vector of dengue fever, has been described by Hawley (1988). The distribution included most of Asia north to Beijing in summer, the eastern United States from Iowa, Wisconsin to the Mexican border, eastern Brazil, islands in the Indian Ocean including Madagascar, Fiji in the Pacific and, most recently, Genoa, Italy (Parodi 1991). The possible global distribution of this important species, under (a) present climatic conditions and (b) the postulated climate change, is shown in Fig. 4, which was again produced using the CLIMEX model. Comparison of Fig. 4 with the present distribution indicates that the species poses a threat to a much wider area than it presently occupies. Given the history of translocation of species around the world (Gillett 1989, Raymond et al 1991), it is only a matter of time before this potential is realized.

Aedes aegypti also breeds in fresh water, in holes in trees, water tanks, tyres, ditches, etc. It is a vector of yellow fever and dengue fever, for which transmission increases with temperature because replication rates increase. Caution is needed in inferring the climatic requirements of *Ae. aegypti* from its observed distribution because it appears to be restricted in some areas by competitive displacement by *Ae. albopictus* (Nasci et al 1989). The latter species appears to make use of a wider range of breeding sites, which enables it to breed more rapidly and so swamp *Ae. aegypti* populations where the two occur together. Reproductive interference is one of the strongest influences on species' distributions and results in highly unstable populations.

It is expected that climatic warming will be accompanied by a disproportionate increase in the frequency of extreme events such as floods, droughts and heat-waves (Wigley 1985). Such events often cause epidemics of disesase, by creating conditions which enable high rates of transmission in areas which normally have low rates of transmission. Epidemics occur because of sporadic transmission in certain age groups who have missed infection for several years and so failed to acquire immunity before accelerated transmission resumes. Epizootics of Rift Valley fever in east Africa are a good example; there were outbreaks in 1951–1953, 1961–1963, 1967–1968 and 1977–1979 associated with above average rainfall (Davies et al 1985). Outbreaks of Murray Valley encephalitis in Australia were associated with two successive years of above average rainfall (Forbes 1978).

Surface water and pollution associated with water management and urbanization

Irrigation is likely to spread and practices are likely to change in response to climate change, with malaria and Japanese encephalitis being the diseases most likely to be affected. The outcome of such changes is very uncertain, because some vectors will benefit while others will suffer from the changes. For example, Amerasinghe et al (1991) describe contrasting changes in the relative abundance of different species of mosquito following the establishment of a large irrigation scheme in Sri Lanka. Similar observations have been made in other countries (WHO 1990).

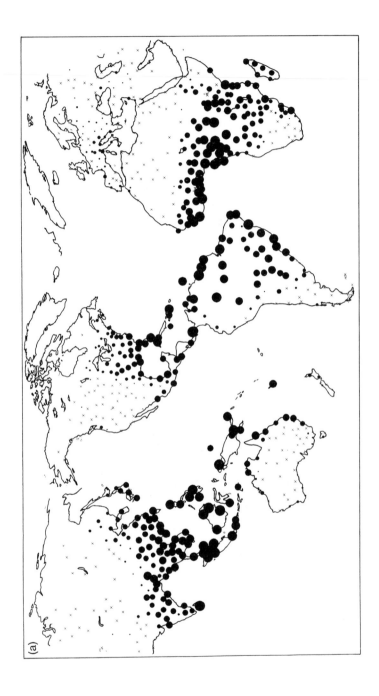

(a)

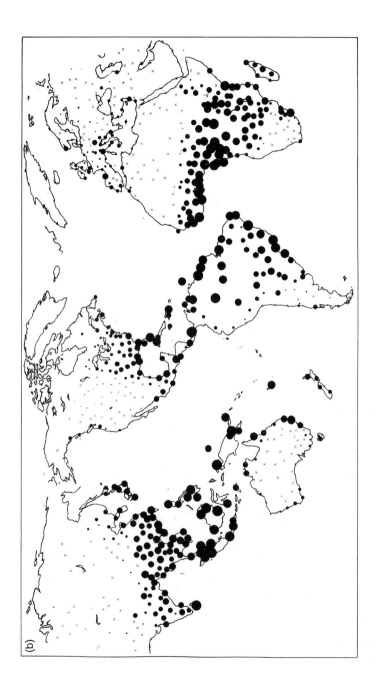

FIG. 2. The geographical distribution of climates suitable for malaria transmission estimated using CLIMEX, (a) based on the present climate and (b) under the climate change scenario described in the legend to Fig. 1. The size of the circles is proportional to the favourableness of the climate (CLIMEX growth index, GI) for seasonal transmission of malaria and crosses indicate places where GI = 0. Figure 3 is an enlargement of the European area of Fig. 2.

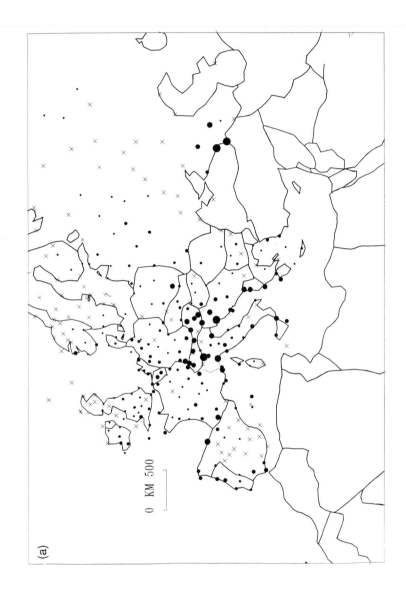

(a)

0 KM 500

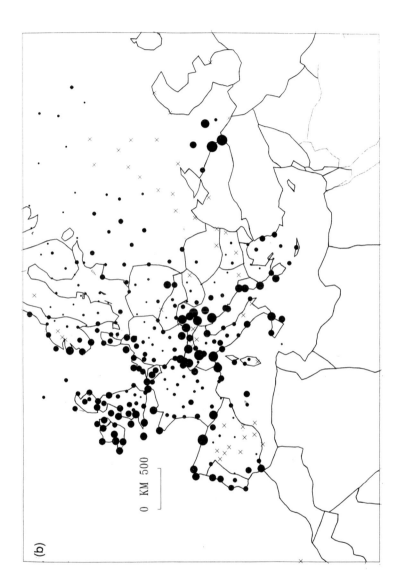

(b)

0   KM   500

FIG. 3.   The geographical distribution within Europe of climates suitable for malaria transmission estimated using CLIMEX, (a) based on the present climate and (b) under the climate change scenario described in the legend to Fig. 1. This figure is expanded from Fig. 2 (see legend for further details).

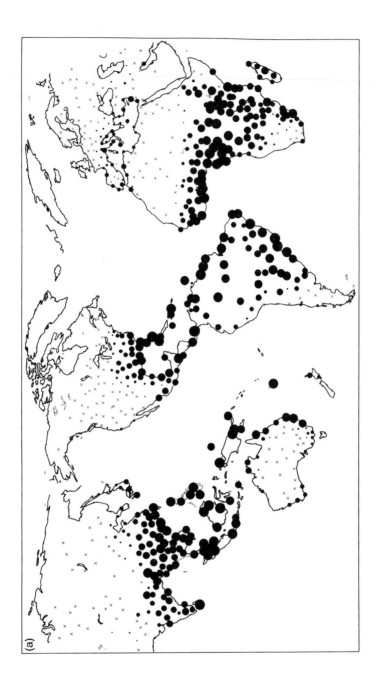

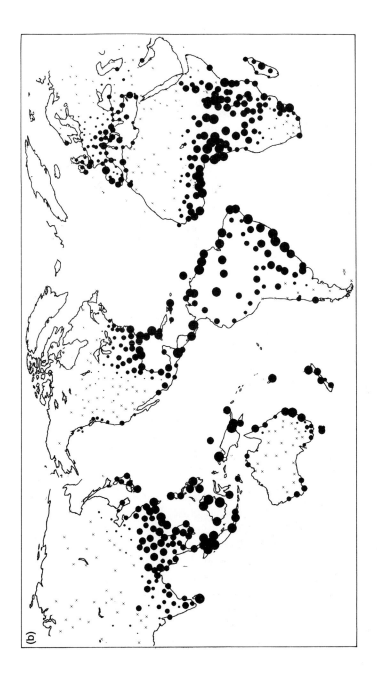

FIG. 4. The geographical distribution of climates suitable for permanent persistence of *Aedes albopictus* estimated using CLIMEX, (a) based on present climatic conditions and (b) under the climate change scenario described in legend to Fig. 1. The size of the circles is proportional to the favourableness of the climate (CLIMEX ecoclimatic index, EI) for permanent persistence of the mosquito. Crosses indicate places where climate is not suitable for the mosquito to live (EI = 0).

Irrigated ricelands are an important breeding site for many species of mosquitoes which are both pests in their own right and vectors of important diseases. With increasing demand for rice to feed Asia's growing population the production of rice will need to increase, and this increased production could involve more irrigation.

Projected sea-level rises of 0.2–0.4 m over the next few decades will cause inundation of some low-lying areas and so alter the distribution of the breeding sites of salt water-breeding mosquitoes.

Urbanization increases the concentration of human hosts and domestic animals and has created new breeding sites in sewage and polluted water for mosquitoes such as *Culex quinquefasciatus*, a vector of lymphatic filariasis (Service 1989). This mosquito has increased in numbers throughout the tropics in response to the inability of developing countries to provide adequate waste water drainage and sewerage in the face of burgeoning urbanization. It has been suggested that this species is more tolerant than other species of pollutants such as detergents and also develops resistance to insecticides very readily. Increased breeding of both *Culex quinquefasciatus* and *Culex pipiens* in waste water contributes to the transmission of arboviruses such as those causing St Louis encephalitis in the USA and Rift Valley fever in Africa.

Human activity has also provided abundant freshwater breeding sites for vectors of dengue virus and yellow fever virus. *Ae. aegypti* and *Ae. albopictus* are both widespread and can breed in any small container of fresh water, such as roof gutters, tins or flower pots. However, the biggest problem is associated with inadequate water supplies and the resultant need to store water in pots and tanks.

Vegetation clearing or re-encroachment

The incidence of lymphatic filariasis, caused by the worms *Wuchereria bancrofti* and *Brugia malayi*, is greatly reduced by environmental development. A much lower incidence has been found in highly populated islands in Indonesia than in rural areas (Lim 1986). This reduction in filariasis is associated partly with destruction of the rainforest canopy which shades the breeding sites of the vector mosquitoes and partly to the different behaviour of humans who no longer enter the forests after clearing a living area.

An example of expansion of malaria into a malaria-free habitat, following forest clearing, is provided by Matola et al (1987). They recorded sporozoite-positive specimens amongst abundant *Anopheles funestus* and *Anopheles gambiae* spp. in the Usambara mountains in Tanzania in the 1970s and 1980s. This followed extensive clearance of the rainforest for agriculture and movement of a large number of infected immigrants to the mountains from the surrounding lowlands where there is intense malaria transmission. Average temperatures increased from 12.8–15.6 °C to 17.7–20.8 °C after clearing of the primary forest. Apart from increasing temperatures, forest clearing also opens up bodies of water to direct sunlight, which facilitates the breeding of *An. gambiae* in Africa, but eliminates *Anopheles leucosphyrus* breeding in Indonesia.

Deforestation in South America has produced man-made savannahs, with a resultant adaptation of triatomine bugs to human dwellings. Normally forest-dwelling insects, the bugs attack forest invaders who become accidental hosts during the early stages of colonization of rainforest. The bugs then adapt to human houses when the forest is cleared and endemic cycles of Chagas' disease develop within the human and domestic animal populations (Forattini 1989). Forest clearing has also led to the creation of localized breeding populations of the sand-fly *Lutzomyia longipalpis*, a vector of visceral leishmaniasis (Kala-azar), and has encouraged breeding of foxes and dogs, which are an alternative host for the pathogen in Latin America (Lainson 1989).

The screw-worm flies *Cochliomyia hominivorax* and *Chrysomyia bezziana*, which cause human and animal myiasis, and also the tsetse fly are very dependent on vegetation cover. Clearance of forest or bush is an accepted method of control of these flies. Re-encroachment of vegetation, in response to increased rainfall with climate change or to demographic changes in human or livestock populations, will result in a resurgence of these pests. The acquired immune deficiency syndrome (AIDS) may affect the population density of humans in some parts of Africa already affected by malaria. It has the potential to massively reduce population densities in rural areas. If such human debility and the associated breakdown of administrative systems occurs, the vegetation cover that the tsetse fly requires is likely to revive; this would add to the continent's problems by causing a resurgence of trypanosomiasis.

Human and animal population densities and movements

The increasing density of human populations is conducive to an increase in the transmission rates of both vector-borne and directly transmitted diseases. Rates of transmission depend largely on the standard of living of the human population, with poverty and poor housing encouraging the propagation of vectors such as body lice, fleas, triatomine bugs and those mosquitoes which breed in standing water. With the projected increase of human populations in a world with limited resources, we are likely to see a global resurgence of many diseases, including bubonic plague, Chagas' disease, malaria and louse-borne typhus. A resurgence of dengue fever is already evident.

Susceptibility to pesticides

It is easy to overlook an environmental change which poses one of the greatest threats to sustainable control of vector-borne diseases, namely the widespread use of pesticides. Repeated exposure of vectors to pesticides used in agriculture and public health results in the permanent loss of one of the most valuable of the world's resources—susceptibility of pests to pesticides. As well as providing breeding sites, irrigation for agriculture is also claimed to accelerate the selection of resistant strains of mosquitoes (Lacey & Lacey 1990) as a side effect of intensified spraying of improved varieties of rice with pesticides. Control of malaria, by the reduction of vector populations, is becoming increasingly difficult as a result of these agricultural practices.

Research needs

Given the high degree of uncertainty about the nature of climate change, the most appropriate strategy is to develop analytical tools which will allow us to evaluate a wide range of scenarios. Effective responses to the impacts of environmental change will require reliable and readily accessible information on the environmental requirements of different vector species. Given the numerous species of vectors concerned, there is little prospect of gaining detailed ecological data on every one. A practical strategy is to make the most of available information and concentrate effort on improving information in areas where there is likely to be the greatest need. Different analytical and predictive tools are needed for different species of vector. Where the data are available, realistic, climate-driven population models are the best to use. For most species, models which capture the essentials of the species' response to its environment will be all that is feasible. Generic, climate-matching models, such as CLIMEX, have a major role to play in such cases.

Intelligent databases incorporating interpretive facilities are needed in order to provide information in a readily available form. These will be more effective if they are user-friendly, global in scope, available through computer networks, can be extended readily and are within the reach of developing countries.

Research in developed countries should also be concentrated on: the living conditions of the poor, in particular on the protection from mosquitoes offered by housing design; the impact of climate change on intermediate hosts, such as white-footed mouse and white-tailed deer in the case of Lyme disease; the impact of climate change on human activity in relation to exposure to vectors; and on how agricultural patterns and forest growth and type affect the presence of vectors and hosts.

Response strategies

The world community is fully stretched in its present efforts to maintain the *status quo* in the case of malaria, and to make significant improvements in public health in developing countries. The prospect of widespread environmental change, particularly urbanization and the loss of susceptibility to pesticides, combined with a decline in wealth is daunting indeed. The development of sound risk assessment and planning procedures, based on community vulnerability, environmental receptivity to pathogen transmission and vigilance of health services (Birley 1989), is needed urgently. New technology, such as remote sensing, is providing tools with which to identify foci of transmission and thus local areas for treatment

(Linthicum et al 1991). However, caution is needed because experience with major development projects has not been good. It indicates that the most important requirement is a suitable management organization for each project (Ackers & Smith 1988). The preference is strongly for small projects rather than massive ones which displace the poor and greatly modify the environment.

Control of vector breeding sites is central to the suppression of diseases such as malaria and dengue fever in cities and Chagas' disease in rural villages. Poverty is the main constraint and this is caused by those factors that the world community is least willing to tackle, namely unsustainable human population growth, degraded natural resources, unfair economic arrangements between developing and developed countries, and the widening gap between the rich and poor.

If the history of international movements of some very important vectors such as *Culex pipiens* and *Ae. albopictus* is anything to go by, the next century is likely to see a rapid colonization of all suitable geographical areas by such species. This raises the question of the lack of rigour shown by quarantine authorities with regard to public health issues, particularly in developing countries. Similar laxity is evident in relation to the potential impacts on the natural environment of species translocations (Sutherst 1993). Improved quarantine surveillance, based on better risk assessments (Sutherst et al 1991), is sorely needed in the face of rapidly increasing international traffic.

Conclusions

If global warming does occur, it will lead to a redistribution of many vector-borne diseases. National and international efforts will be needed to manage the effects of the spread. National health authorities should be adequately prepared to handle the control of vector-borne diseases through vector control, vaccination of individuals at risk and drug treatment (WHO 1990). Climate change will also affect vector populations. Increased temperatures will accelerate the development of the pathogens carried by the vectors, and so accelerate disease transmission. While the details of the changes in vector populations that will result from environmental changes remain unclear, the best response strategy is to attack the causes of those environmental changes. Models then need to be further developed to allow us to anticipate and prepare for a change in the status of each disease. While human populations continue to increase there is little prospect of an improvement in the status of vector-borne diseases. It is much more likely that there will be a great decline in the success rates already achieved, accelerated by the widespread development of resistance to pesticides and drugs for which there are few replacements being developed.

Acknowledgements

Many people were helpful in supplying me with information for this paper. In particular, Dr C. F. Curtis and Dr B. Kay gave both information and advice which was useful. Mr G. F. Maywald assisted with the computing.

References

Ackers GL, Smith DH 1988 Design and management of development projects to avoid health hazards. J Trop Med Hyg 91:115–129

Amerasinghe FP, Amerasinghe PH, Malik Pieris JS, Wirtz RA 1991 Anopheline ecology and malaria infection during the irrigation development of an area of the Mahaweli project, Sri Lanka. Am J Trop Med Hyg 45:226–235

Birley M 1989 Guidelines for forecasting the vector-borne diseases: implications of water resource development. WHO, Geneva (PEEM Guidelines Ser 2)

Davies FG, Linthicum KJ, James AD 1985 Rainfall and epizootic Rift Valley fever. Bull WHO 63:941–943

de Zulueta J, Ramsdale CD, Coluzzi M 1975 Receptivity to malaria in Europe. Bull WHO 52:109–111

Dutta HM, Dutt AK 1978 Malarial ecology: a global perspective. Soc Sci & Med 12:69–84

Forattini OP 1989 Chagas' disease and human behaviour. In: Service MH (ed) Demography and vector-borne diseases. CRC Press, Boca Raton, Fl, p 107–120

Forbes JA 1978 Murray Valley encephalitis 1974 and the epidemic variance since 1914 and predisposing rainfall patterns. Australasian Medical Publishing, Glebe, New South Wales

Gillett JD 1989 The maintenance and spread of insect-borne disease by the agency of man. In: Service MH (ed) Demography and vector-borne diseases. CRC Press, Boca Raton, Fl, p 35–46

Hawley WA 1988 The biology of *Aedes albopictus*. J Am Mosq Control Assoc (suppl 1) 4:1–39

Hoogstraal H 1981 Changing patterns of tickborne diseases in modern society. Annu Rev Entomol 26:75–99

Lacey LA, Lacey CM 1990 The medical importance of riceland mosquitoes and their control using alternatives to chemical insecticides. J Am Mosq Control Assoc (suppl 2) 6:1–93

Lainson R 1989 Demographic changes and their influence on the epidemiology of the American leishmaniases. In: Service MH (ed) Demography and vector-borne diseases. CRC Press, Boca Raton, Fl, p 85–106

Lim BL 1986 Filariasis in Indonesia: a summary of published information from 1970–1984. Tropical Biomed 3:193–210

Lines JD, Wilkes TJ, Lyimo EO 1991 Human malaria infectiousness measured by age-specific sporozoite rates in *Anopheles gambiae*. Parasitology 102:167–177

Linthicum KJ, Bailey CL, Tucker CJ et al 1991 Towards real-time prediction of Rift Valley fever epidemics in Africa. Prev Vet Med 11:325–334

Matola YG, White GB, Magayuka SA 1987 The changed pattern of malaria endemicity and transmission at Amani in the eastern Usumbara mountains, north-eastern Tanzania. J Trop Med Hyg 90:127–134

McDade JE, Newhouse VF 1986 Natural history of *Rickettsia rickettsii*. Annu Rev Microbiol 40:287–309

Nasci RS, Hare SY, Willis FS 1989 Interspecific mating between Louisiana strains of *Aedes albopictus* and *Aedes aegypti* in the field and laboratory. J Am Mosq Control Assoc 5:416–421

Parodi P 1991 Aggiornamento su *Aedes albopictus*. Nuovo Prog Vet 46:272–273

Pittock AB, Nix HA 1986 The effect of changing climate on Australian biomass production—a preliminary study. Clim Change 8:243–255

Raymond M, Callaghan A, Fort P, Pasteur N 1991 Worldwide migration of amplified insecticide resistance genes in mosquitoes. Nature 350:151–153

Service MH (ed) 1989 Demography and vector-borne diseases. CRC Press, Boca Raton, Fl
Sutherst RW 1991 Pest risk analysis and the greenhouse effect. Rev Agric Entomol
    79:1177–1187
Sutherst RW 1993 The potential advance of pests in natural ecosystems under climate
    change—implications for protected area planning and management. Academic Press,
    London, in press
Sutherst RW, Maywald GF 1985 A computerised system for matching climates in ecology.
    Agric Ecosyst & Environ 13:281–299
Sutherst RW, Maywald GF, Bottomley W 1991 From CLIMEX to PESKY, a prototype
    version of a generic expert system for pest risk analysis. (Proc WMO/EPPO/NAPPO
    Symp Pract Appl Agrometeorol Plant Prot). Bull OEPP (Organ Eur Mediterr Prot
    Plant) 21:595–608
Weihe WH 1991 Human well-being, diseases and climate. In: Jäger J, Ferguson HL
    (eds) Climate change: science, impacts and policy. Cambridge University Press,
    Cambridge (Proc Sec World Clim Conf) p 352–359
WHO 1990 Potential health effects of climatic change. WHO, Geneva (Rep WHO/PEP/
    90/10)
Wigley TML 1985 Impact of extreme events. Nature 316:106–107

DISCUSSION

Hoffmann: What is the current role of pesticides in the fight against disease vectors?

Sutherst: There is considerable effort to move away from the use of pesticides in all pest management areas. There have been some tremendous successes with the tsetse fly in Africa using traps baited with cattle urine. Herdsmen collect cattle urine and put it under a little cloth tent, and tsetse flies are attracted to the urine and get caught in the trap. These traps are really effective and have depopulated areas of tsetse flies (Dransfield et al 1991).

There are serious problems of resistance in sand-flies and mosquitoes. As I said earlier (p 101, p 137), there are not many chemical pesticides left to use against these disease vectors. It costs more than 50 million dollars to produce each new chemical, and the companies can't afford it. Health regulations, safety regulations and all sorts of often sensible restrictions are preventing the search for new chemicals.

Hoffmann: I used to live in Minneapolis, Minnesota, where there is a big problem with mosquitoes in the summer. The strategy there was to spray the breeding areas with malathion on a regular basis. That effort has kept the mosquito population down and made life more tolerable in the summer.

Sutherst: There are good data from the USA showing that the risk of getting St. Louis encephalitis is highly related simply to whether the houses are screened or not. The standard of housing is very important; some of the recommendations in these areas are that people should become more aware of their behaviour. Is it sensible to eat outside in the garden at dusk when mosquitoes are most active, for example?

Hoffmann: The local laws in California and in most of the USA require every window in homes to have a screen. You can't sell a house if it doesn't have screening.

Sutherst: That is a sensible development.

Bradley: With malaria, the success of the residual insecticides in reducing transmission is followed in some areas by development of resistance. In other places people continue to use DDT (dichlorodiphenyltrichloroethane) in large quantities, where in fact the main vectors rest and bite outside the houses, and where there's no evidence that DDT is doing any good. A large quantity of DDT is used unnecessarily. However, in a lot of places, if the use of DDT is banned because of the general environmental issues and the effects on agricultural crops, a quite effective means of controlling malaria is removed. There is a tendency not to differentiate between the use of residual insecticides for agricultural and for health purposes.

Malaria has got us on the run. The situation has been getting worse and worse over the last few years. The one thing that's looking promising at the moment, much more promising than people had thought, is the impregnation of bed nets with pyrethroid insecticides. This is extremely cheap to do, uses less insecticide, and yet has reduced mortality from malaria in places such as the Gambia where it is very difficult to control malaria.

Zehnder: There are places, Africa for example, where DDT is still used, because there is no choice. Given the choice between dying as a result of DDT poisoning in 20–30 years or getting fever the next morning, everyone would, I think, spray DDT. There is a need for more complex management systems.

Bradley: In fact, DDT actually isn't used a lot in Africa because it doesn't do much good, simply because there's so much transmission that even after spraying there are still enough mosquitoes around to continue to transmit malaria. It may reduce the mosquitoes, but it doesn't reduce the malaria. DDT is still used in large quantities in Asia.

Sutherst: Could you tell us something about the work that's being done with polystyrene beads in latrines?

Bradley: This is an alternative method of controlling urban mosquitoes, which are mostly culicines, which not only bite people and keep them awake at night, but also transmit filariasis. These mosquitoes breed in dirty water and they live in flooded pit latrines, the common method for safe disposal of excreta. The water table is high in most of the coastal cities of the world, and the pits therefore flood. There can be phenomenal numbers of mosquitoes. The biting rate in Jakarta, for example, is 300 000 bites per annum per person. Small polystyrene beads, which can be transported very cheaply, are boiled up in a cooking pot so that they expand and then are put into the pit latrines where they float on the top, forming a layer several millimetres deep; the mosquitoes cannot get their breathing syphons through this layer of beads. There are pit latrines where this method has been working for seven years without any maintenance costs or any effort at all.

Edwards: One of the forms of protection commonly used in New Zealand and California is insect repellents. The indigenous population go out with rather less than half of the surface of their bodies covered with clothes, and the rest is covered with a thin layer of a highly effective insect repellent. At least 1% of this population is pregnant. Is anything known about the absorption and systemic effects of the chemicals which are often covering the skin of a large number of healthy individuals?

Sutherst: There were reports of brain damage in children whose parents covered them with the insect repellent DEET (*N,N*-diethyl-3-methylbenzamide) every night instead of using nets (Roland et al 1985). In some places the use of topical insecticides has changed the course of wars. Putting pyrethroid insecticides onto soldiers' leggings has allowed them to occupy areas the opposition could not. This happened in Vietnam and in Zimbabwe.

Edwards: Do these chemical repellents work?

Bradley: DEET certainly seems to work and is not very toxic, though I would have reservations about pregnant women using it.

Sutherst: The military have made comparisons of mosquito biting rates with and without repellents in Canada, for example, and have actually learnt quite a lot about the behaviour of the vector at the same time.

Lake: What effects will environmental change have on the major pests of plants and animals, for example, locusts, Colorado beetles and the various flies that infest livestock?

Sutherst: Locust problems are event-driven. There tend to be swarms of locusts after high rainfall in desert areas. If, as has been predicted, there are more extreme events in the future, the locust problem is likely to worsen. Summer rainfall will probably increase whereas winter rainfall will probably decrease. It's quite likely that there will be more high rainfall episodes, leading to increases in outbreaks of locust and army worm infestations, for example. We did some work on Colorado beetle (*Leptinotarsa decemlineata*) using the CLIMEX model about two years ago (Sutherst et al 1991). This suggested that with the predicted change in climate Colorado beetle will be comfortable all over Europe, as far north as Scandanavia. The changes to other pests depend on location. North America seems to be reasonably well protected, as are continental Europe and northern Asia, because the winters will still be so severe that the pests will be able to make only a small push forward into these areas. With Colorado beetle, the problem that now exists in France will probably move further north, to Sweden, for example; the crop will move and the pests will move with it. The problems that are going to occur in temperate areas are already being experienced in the tropics. They are being handled in the tropics at present, so the problem is largely one of adaptation. These changes will certainly increase costs and reduce productivity and profitability, because there will be a greater need for pest control.

The unpredictable aspect is the effect of environmental change on the interaction between natural enemies and the insect pest. The pest might gain a greater advantage from a temperature rise than the biological control agent, the natural enemy. The interaction between them would then be put out of balance.

Zehnder: If it is getting more difficult to produce more new pesticides to overcome resistance, it becomes even more important to optimize control by the natural enemies. Climate change may affect the pests and/or their enemies positively, or, in the worst case, affect the pest positively and the enemies negatively. Certain management control measures may favour natural enemies. Are such aspects built into your models?

Sutherst: The difficulty is that people in the cropping area want to look at two or three thousand species of pests, and the resources to look at every one in detail simply aren't available. With our CLIMEX model we can make rapid projections. For example, the simulation for tsetse fly that I described (p 128) was produced in one morning, whereas other techniques would have taken several weeks. To study interactions between biocontrol agents and pests you need a detailed simulation model of each, and there are few of these for the major pests and biocontrol agents. Now that biological control is becoming more commercialized, the resources to study these sorts of interactions will probably be provided in part by chemical companies, whereas previously the funding came from the growers or government. As we see it, the main need, if we are to adapt to climate change, is for tools with which to analyse these sorts of interactions and their relationship with climate, so that whatever might happen can be taken into account and we can be at least partly prepared.

Zehnder: There is clearly a need for studies of population dynamics.

Bradley: If the predicted changes in temperature do happen, there will probably be a change in the staple crops; the cereals grown will actually change. People are reluctant to change, but in Bangladesh, for example, there has been quite a substantial shift from rice to wheat over the last few years.

Sutherst: Some countries might be forced to do this. In Australia, the sort of climate changes that we have been talking about would virtually wipe out the whole wheat industry. At present, the areas which have a suitable climate are also those with suitable soils and physical characteristics. That climatic 'pocket' will move into hilly country with poor soils that cannot be cultivated. In the other part of the country where we grow wheat, in the west, a drop in the winter rainfall is expected and the pocket of suitable climatic conditions will move out into the sea. There will have to be alternatives and we are going to have to learn to eat different crops. I should emphasize that these sorts of changes will all be superimposed on our struggle to feed the growing population of the world. With all this other disruption as well, the problem is going to be beyond our means unless we can divert vast sums from the military.

Another issue is natural environments. Many natural parks have been established around the world to preserve specific vegetation or animal communities. If the climate that is suitable for those communities moves several hundred kilometres, then the reason for the existence of the national park has gone. The logistics of simply planting the trees somewhere else are beyond us.

Zehnder: You didn't paint us an encouraging picture. As you said, humans are the biggest pest, but I think we humans are also inventive and good at figuring out ways to overcome obstacles and problems.

Sutherst: I think you can only say we are managing at the moment if you use what in other countries is called the one-eyed trick, shutting your eyes to all the problems that are not being coped with. The predicted 5.4 billion population in the southern hemisphere is going to test us. I'm sure we have the capacity and the intellect to manage; the question is one of priorities, putting effort into these areas instead of building television stations!

References

Dransfield RD, Williams BG, Brightwell R 1991 Control of tsetse flies and trypanosomiasis: myth or reality? Parasitol Today 7:287–291

Roland EH, Jan JE, Rigg JM 1985 Toxic encephalopathy in a child after a brief exposure to insect repellents. Can Med Assoc J 132:155–156

Sutherst RW, Maywald GF, Bottomley W 1991 From CLIMEX to PESKY, a generic expert system for pest risk assessment. Bull OEPP (Organ Eur Mediterr Prot Plant) 21:595–608

Human tropical diseases in a changing environment

David J. Bradley

Department of Epidemiology & Population Sciences and Ross Institute, London School of Hygiene and Tropical Medicine, Keppel Street, Gower Street, London WC1E 7HT, UK

Abstract. A taxonomy of the human environment and of its components is set out as a basis for understanding the complex determinants of human diseases in the tropics. The scale, nature and trends of tropical diseases are related to changes in environmental components, life systems and habitats, and these are analysed, with emphasis on vector-borne diseases, especially malaria. Although it is feasible to predict the consequences of changing single climatic variables for specific parts of parasite and vector life cycles, the probable effects of simultaneously changing several variables are far more difficult to predict because many qualitative and quantitative determinants of disease are likely to be affected and the effects of covariance are poorly understood. Local and specific changes are likely to overshadow more global climatic changes. The effects of global temperature rise are explored, with particular emphasis on the basic case reproduction rate for two species of human malaria parasites. Even a 2 °C temperature rise greatly increases this rate at the lower end of the temperature range compatible with transmission. The importance of secondary vectors may increase. Implications for transmission in Europe and for imported tropical diseases are discussed.

1993 Environmental change and human health. Wiley, Chichester (Ciba Foundation Symposium 175) p 146–170

In any consideration of the effects of environmental change on tropical diseases, at a meeting especially concerned with Europe, there are at least three separate but related issues. They concern (i) the effects of environmental change on the health of tropical populations of developing countries, (ii) the consequences of these and other changes for the level of tropical diseases imported into Europe and (iii) the effects of climatic changes on the distribution of tropical diseases and whether they spread towards or within Europe. There can be little doubt that the first category will be of the greatest importance to the world and even to Europe.

Although the formal definition of health has been much debated, there is a substantial measure of agreement on what health is. By contrast, although most writers on the environment are clear what *they* understand by the term, there is very great divergence of view: some think of wilderness, others of pollution and others of habitat destruction or of water supply. Discussion of environment and health has suffered because people have started from narrow and partial definitions of the environment. I shall therefore begin with a working approach to defining the human environment.

It is the habit of most workers in the area of international health to apply the words 'tropical' and 'developing' to a country almost interchangeably. This paper is no exception. The logic for such usage is set out in Fig. 1. 88% of the world's low income countries, containing 95.6% of low income country population, lie partly or wholly between the tropics (World Bank 1992; only those countries providing data and with a population exceeding one million are considered). 65.4% of middle income country populations are also at least partly tropical, in comparison with only 3.4% of those from high income countries. Conversely, of those living partly or wholly in the tropics, 79.9% have a low income and only 0.8% have a high income. Only 4.4% of low income country populations live outside the tropics, and Pakistan, Nepal and Afghanistan account for almost all of them.

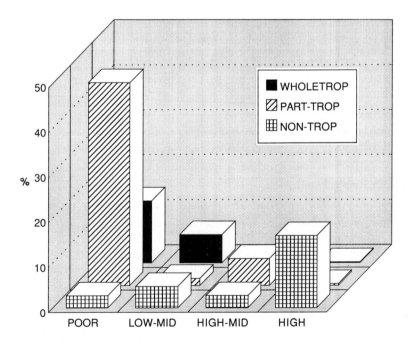

FIG. 1.   The relation between poverty and a tropical location, expressed as the location of national populations. (Data from World Bank 1992.)

What is the environment?

The most appropriate definition of the environment, for the purposes of discussions on health and disease, resembles that put forward by the ecologists Andrewartha & Birch (1954). The environment of a person is all that is other than him or her and which, directly or indirectly, influences or is influenced by that person. Three aspects of this definition are significant — it is comprehensive, it involves people and is anthropocentric, and it refers to the individual. There are real problems in speaking of the environment of a population. To be sure, the environment of many individuals will be very similar to that of their neighbours, but the environment of each individual includes other people. This puts the population issues into their appropriate context.

The broad connotation is also needed. If the environment is defined too narrowly, people are encouraged to give biased attention to it: narrow-minded interventions may give rise to more medium-term problems than they solve in the short term. The person who wishes to stop the use of DDT (dichlorodiphenyltrichloroethane) abruptly in some parts of the tropics where it is used only to control malaria may kill a greater biomass of people than he saves of other organisms.

The concept of the environment is inherently anthropocentric. It is not the 'world' or any total ecosystem. Rather, it carries the idea of interaction between the human being and his or her surroundings.

Most aspects of life and health are environmentally determined, especially in developing countries, so that the most acutely relevant issues concern environmental change in an area over time. Comparisons should also be made between tropical and temperate environments as well as considering the changing nature of each and their consequences for health and vice versa. There are also marked social and economic differences between the populations of tropical and temperate latitudes. The physical and biological environment cannot be treated separately from the economic and socio-cultural environment without distortion.

Health in tropical countries

There is a continuous series of countries from the very poor, such as Mozambique with a GNP per head of £42 annually, to the very affluent, such as Switzerland with one of £17 000 per year. The main countries of Africa and south Asia have notably low incomes, and their health patterns are similar, though the scene is worse in Africa. Predominant disease categories are gastrointestinal and respiratory infections. These are the principal cause of death in children. The infant mortality rate is very high, and life expectancy low. The commonest single specific cause of death in adults is pulmonary tuberculosis, but trauma is also very common and AIDS (acquired immunodeficiency syndrome) is of rapidly increasing importance. In other words, the main causes of mortality are the diseases of poverty rather than the classical tropical diseases. Of these, only malaria figures as a major killer, particularly in Africa where it ranks with diarrhoeal and respiratory infections as the leading cause of death in children and adults in, for example, Tanzania (UNICEF 1985, Bradley 1991) and The Gambia (McGregor 1991).

In developing countries, infections are the main causes of disease and death, and the poorer the country, the more they predominate. Malnutrition and trauma are of great importance also, and it is difficult to disentangle the respective roles of malnutrition and infection, particularly in young children. The chief environmental determinants of infections are the complex of factors included in the term 'poverty', and the warm climate. The most direct effect of temperature is on the transmission of the infections caused by animal parasites (protozoa and helminths) and arboviruses. Almost all parasites (used here in this limited sense of worms and protozoa) have stages in their life cycle outside the human host which are temperature dependent. They take place in insects, molluscs and even in the soil, and depend on the ambient temperature, which needs to be high enough to allow development to take place. Usually, the higher the temperature, the shorter the part of the cycle outside man (the 'extrinsic cycle') and hence the more rapid the transmission.

The environmental determinants of these health hazards are diverse, but many, especially those of the classical 'tropical' diseases, are related to water and to human excreta as well as to the ambient temperature and humidity. Those infections related to water may be classified into four main groups.

Water-borne diseases, where the pathogen is transmitted by ingestion of contaminated water. Cholera and typhoid are often spread this way.
Water-washed diseases involve faecal–oral or other spread from one person to another facilitated by the lack of adequate supplies of water for washing. Many diarrhoeal diseases as well as infections of the skin and eyes are transmitted in this way.
Water-based infections are those caused by pathogenic organisms which spend part of their life cycle in aquatic organisms. The schistosomes and other trematode parasites which parasitize snails, and guinea worm (dracunculiasis), which is spread through minute aquatic crustaceans, are examples.
Water-related insect vectors include those which breed in water, such as the mosquitoes which spread malaria and filariasis, arthropods which carry viruses such as those causing dengue and yellow fever, and black-flies which transmit river blindness. Also, some of the tsetse flies that transmit sleeping sickness bite preferentially near water.

The consequences of environmental change

The array of environmental changes affecting tropical countries is vast. For our purposes they fall into two broad categories: those changes deliberately instigated in order to improve health, and those where changes to health are a side-effect of environmental alterations produced for some other purpose.

The first group consists of improvements in domestic water and sanitation, improved occupational hygiene and solid waste collection, etc., and will not be considered at length here. The other environmental alterations may make health worse, in at least two ways. The environmental change may have a direct effect on disease transmission or causation: irrigation, for example, may increase habitats for snail intermediate hosts or breeding places for mosquitoes. Also, it may influence the distribution of disposable income: many economic development projects which have an aggregate beneficial effect in fact make the better-off groups richer but make the poorest even poorer than before. At the margins of subsistence, any reduction in an already grossly inadequate income will be reflected in malnutrition and disease.

A changing environment for a population may either result from actual changes to that environment or result from migration of the population to a new site, and hence a new environment. The numbers of people moving to growing cities, and the number of migrant agricultural labourers and refugees, make this form of environmental change a major source of ill-health. Among the classical tropical diseases, malaria is a particular hazard for migrants from overcrowded hills to malarious plains; the disease also manifests itself in infected and hungry people when they are re-fed. 'Malaria of the tropical aggregation of labour' is an accepted technical term.

The prefix 'global' is often added to environmental change without careful thought. Environmental changes are the result of human activities that are unevenly distributed over the world, and in the vast majority of cases the effects on health are most felt at the site of the changes, though the driving forces for the environmental changes, often human cupidity and the structure of markets, may originate far away. Even in the case of climatic changes caused by the 'greenhouse' gases, where there is a global effect, the resulting changes in weather will vary greatly at different sites (see Table 1). Moreover, there is every reason to believe that the consequences for Europe of environmental changes occurring in and affecting developing (and other) countries locally are likely to far exceed the consequences of truly global changes so far as life generally is concerned, and even in relation to tropical diseases specifically.

Environmental changes can be grouped into four broad formal types — changes of ambience, habitat, the life system, and in components of the habitat. Table 2 shows how these categories relate to those of Andrewartha & Birch (1954) and to the major changes taking place in tropical countries. In addition to climatic change (ambience), the major changes affect the habitat and the life system related to it. Perhaps the key changes, and certainly those most apparent, are water resource developments, land resource changes (I hesitate to call them development), among which deforestation is prominent, and urbanization, a term that encompasses a great complex of environmental changes for the immigrant and for the child born into the city instead of the countryside. The scale of these changes is very great. All three have major effects on infections

TABLE 1 Predicted changes in regional climates by the year 2030

Place	Temperature change (°C) Winter	Summer	Precipitation change (%) Winter	Summer	Soil moisture change (%) Winter	Summer
Central N. America	+2 to +4	+2 to +3	0 to +15	−5 to −10		−15 to −20
South Asia		+1 to +2	0	+5 to +15	0	+5 to +10
Sahel		+1 to +3	Up but variable		A little down but variable	
South Europe	+2	+2 to +3	Up	−5 to −15	−15 to −25	Variable
Australia	+2	+1 to +2	0	+10	Variable	Variable
Overall	+1.8					
Assume	+2			+5	Little change	

TABLE 2   Alternative categories of the environment

Andrewartha & Birch (1954)	Bradley	Changing aspects

Weather	Ambience	Climate
		  Temperature
		  Precipitation
		  Humidity
		  Surface waters
		  Salinity

Place to live	Habitat	Resource development
		  Irrigation
		  Deforestation
		  Urbanization

Food	Life system	Farming systems Employment

Other organisms
  Same species
  Different species
  Free-living
  Pathogens
	Components	Vectors
		Zoonotic reservoirs
		Infections
		
		Pollutants
		Toxins

and particularly on vector-borne diseases, as do changes in the components of the environment, especially use of insecticides and anti-parasitic drugs and the types of housing (whether viewed as habitat or components of the environment). Changes in life systems are more subtle. A switch from subsistence to cash crops does more than affect an environmental component; it modifies the whole agricultural work pattern and its products. It may well lead on to changes in nutrition and the distribution of disposable income, as would the adoption of a high-yielding, high-input crop variety with reduced drought tolerance.

Consequences of climatic change for vector-borne diseases

The consequences of environmental changes of various types for disease in developing countries are addressed elsewhere in this volume (Bradley 1993) in broad terms. Here, the focus is on the classical vector-borne diseases of the tropics and the effects of global climatic change. Some effects of environmental and habitat change on vector-borne disease are very place-specific, depending upon the locally prevalent vectors and upon potential vectors that can exploit the changed environment. Others, and particularly the effects of rising temperature, are more general; these are explored in this paper, whilst the more specific effects are described in Bradley (1993; this volume). The effects of temperature are explored with emphasis on the most important and best-understood vector-borne tropical disease: malaria.

It has been widely suggested that one consequence of global warming may be a spread of what are now thought of as tropical diseases into temperate regions. This, in simple terms, is plausible, and it is easy to imagine the boundaries of malaria transmission spreading up tropical mountains and along towards the higher latitudes. The two main components of global climatic change are temperature and precipitation, which affects both atmospheric humidity and the extent, depth and flow of surface water. A change in the temperature will have a direct effect on the malaria parasites. Both temperature and precipitation will affect transmission indirectly, through effects on the ecology and behaviour of anopheline mosquitoes and people. It is useful to base the discussion around the predicted changes in climate shown in Table 1, derived from the report of the Intergovernmental Panel on Climate Change (IPCC) (Houghton et al 1990). In particular, I have worked on the assumption that there will be an increase of temperature of 2 °C by the year 2030, accompanied by an average 5% increase in precipitation but with great regional variation such that there will be a fall in precipitation in some areas.

There are four human malaria parasites, which are obligatory parasites of red cells; one of them, *Plasmodium falciparum*, is more important than the others because it is the cause of malignant malaria. A red cell infected with *P. falciparum* has on its surface many small knobs which stick to the capillaries of the brain, slowing blood movement. The infected person may go into a coma, possibly because of local concentrations of tumour necrosis factor or because the oxygen in the blood cannot get to the brain, and the small amount that does tends to be consumed by the malaria parasites themselves. If untreated, the coma of cerebral malaria will usually have a fatal outcome.

Malaria is the most important vector-borne disease of human beings. Estimates from the World Health Organization (WHO) suggest that the number of people exposed to the risk of malaria is of the order of two thousand million; 270 million (probably a more accurate figure) are infected each year, of which about half become ill, and a million or two probably die every year. It is a major public health problem.

Our understanding of the interaction of the environment with malaria is linked to the discovery, made just over 90 years ago, that the disease is transmitted by anopheline mosquitoes. To understand the effects of environment on malaria requires some quantitative epidemiology. The best way to measure the spread of malaria uses the basic case reproduction rate (R_o). This is quite a simple concept: if somebody has malaria and other people are not immune in that particular place, the basic case reproduction rate is the number of people who will catch malaria from that one person after one cycle through the mosquito. For example, with a basic case reproduction rate of two, one person might infect two people, and those two infect four, and gradually more and more people get infected. With a basic case reproduction rate of one half, four people on average infect two before they die or get better; these two on average infect one, so the disease gradually dies out. In places with epidemic malaria R_o fluctuates about one, rising perhaps to three or five; in areas of high endemic malaria such as the West African savannah, it may exceed 1000. This is a very simple and useful concept. Basically, if R_o exceeds one the disease will spread indefinitely; if it is below one the disease will die out. This enables us to begin to simplify the epidemiology of malaria.

The parasites are either in people or in anopheline mosquitoes, and we can reduce the parasitologist's complex picture of malaria to a basic cycle of transmission between man and mosquito. Thus, for our purposes, the considerations are whether the mosquito is susceptible in the first place, and, if it is, how long it lives, how many of its kind there are, how often it feeds, and whether it feeds on man. The crucial point in understanding environment and transmission is that after the mosquito hatches out from a pupa it feeds from time to time, and if it feeds on a person who already has gametocytes of malaria parasites in the blood, the mosquito may get infected. The parasite then develops in the mosquito in what is termed the extrinsic cycle. Because the mosquito is roughly at the ambient temperature of its micro-habitat, the duration of the extrinsic cycle is determined by that temperature. It also depends on the species of malaria parasite (Table 3), but much less on the anopheline species involved.

The duration of the extrinsic cycle is usually about 10 days or more, and until the end of this period the mosquito is not infectious; after this period it is infectious for the rest of its life. Because 10 days is a long time in the life of a mosquito, malaria is transmitted basically by geriatric mosquitoes, those that have bitten and have lived at least 10 days. The entire theory of transmission depends on that. It is also a feature of all other vector-borne diseases; with schistosomiasis, there is a latent period of about a month between the time when the snail is infected and the point at which it becomes infectious to people. The survival of the vector through this period is particularly important. For example, if a thousand long-lived mosquitoes bite people with malaria parasites, after 12 days about 500 may still be left alive and infectious, whereas with a short-lived

TABLE 3 Effects of temperature on the extrinsic cycle of malaria parasites

Plasmodium species	M^a	m^b (°C)
P. vivax	105	14.5
P. falciparum	111	16.0
P. malariae	144	16.0

[a]M is a constant for the species of parasite.
[b]m is the lowest temperature at which the extrinsic cycle can be completed.
Data from Molineaux (1990).

mosquito, there may be only a few remaining alive after the duration of the extrinsic cycle, and these will live for only a short time thereafter. One can therefore estimate the rate of spread of the malaria parasite using the basic malaria transmission equation to calculate R_o, as shown in (1), where a,

$$R_o = \frac{da^2bp^n}{-r\log_e p} \tag{1}$$

represents the probability of the female mosquito biting man in one day; d, the female mosquito density per person; p, the chance of a mosquito surviving through one day; n, the duration of the extrinsic cycle (measured in days); b, human resistance to infection; r, the recovery rate of people from malaria; and e, the base of natural logarithms. The most direct effect of temperature is on n, and $R_o \propto p^n$.

Figure 2 shows the dependence of n upon temperature in the conventional way (A) and in a more convenient form (B). If a climatic change were to affect only n, then the effect of a 2 °C rise in temperature would be to significantly reduce n and raise p^n at lower temperatures. The relation of p^n to temperature for long-lived (high value of p) and short-lived (low value of p) mosquitoes is shown in Fig. 3 for two species of malaria parasite. The effect of a 2 °C rise in ambient temperature on the basic case reproduction rate can be explored more fully using the formulae that best describe the curves in Fig. 2B.

Consider a rise in temperature of 2 °C from an initial temperature of t °C, for a vector with an expectation of survival through one day of p, for a malaria parasite whose extrinsic cycle lasts n_t days at temperature t, and for which the duration of n has the form shown in (2), where M is a constant for the species

$$n_t = \frac{M}{t - m} \tag{2}$$

of parasite and m is the lowest temperature at which the extrinsic cycle can be completed for that species (Table 3). The ratio of transmission after the temperature rise to that before is y, where y is related to p as shown in (3). More generally, for a rise in temperature of c degrees, y is related to p as shown in (4).

$$y = p^{-\frac{2M}{t^2 - (2m-2)t + m(m-2)}} \tag{3}$$

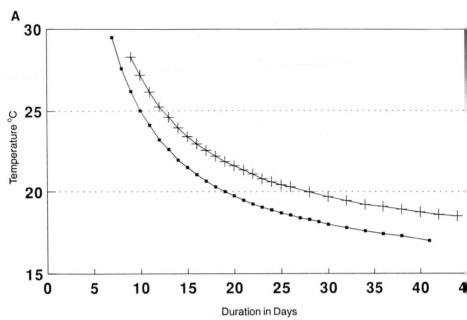

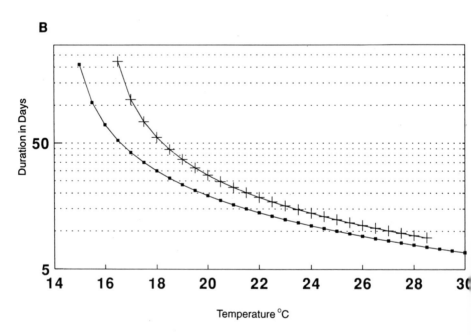

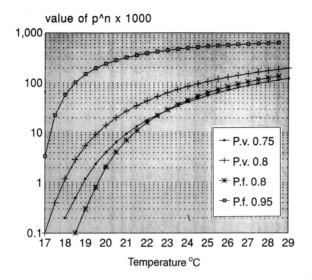

FIG. 3.   Variation of the component p^n ($\times 1000$) of malaria transmission, for selected values of p (the chance of a mosquito surviving through one day), as affected by the change of n (the duration of the extrinsic cycle) with temperature. Values are shown for two species of malaria parasite: *Plasmodium vivax* (P.v.) at $p = 0.75$ (■) and $p = 0.8$ (+) and *P. falciparum* (P.f.) at $p = 0.8$ (✳) and $p = 0.95$ (▨).

$$y = p^{-\dfrac{cM}{t^2 - (2m - 2)t + m(m - 2)}} \tag{4}$$

From $t = m - c$ to $t = m$ the ratio will be infinity because transmission can only occur at the higher temperature. Below $t = m - c$ there will be no transmission at either temperature.

This simplifies for a 2 °C rise in *Plasmodium vivax* (v) and *P. falciparum* (f) as shown in (5). In Fig. 4 these values of y are plotted for vectors of *P. vivax* with values for p of 0.75 and 0.8, and for *P. falciparum* with values for p of 0.8 and 0.95.

$$y_v = p^{\dfrac{-210}{t^2 - 27t + 181.25}} \qquad y_f = p^{\dfrac{-222}{t^2 - 30t + 224}} \tag{5}$$

FIG. 2. (*opposite*)   The effect of temperature on the duration of the extrinsic cycle of the malaria parasites *Plasmodium vivax* (◆) and *P. falciparum* (+). Part A follows the conventional layout (Macdonald 1957), with temperature plotted against the calculated duration of the extrinsic cycle, and Part B uses a layout more compatible with the analysis here.

It is clear that at low temperatures the increase in the basic case reproduction rate caused by a 2 °C temperature rise can be very great. The extreme values are probably of little practical relevance, because the chance of a mosquito surviving for the duration of the extrinsic cycle becomes remote when it is so long. In rather less extreme circumstances, where the extrinsic cycle lasts well under a month, the effect of a temperature rise is still substantial and could well carry R_o above one in areas where previously it had been continuously below one and which were therefore non-malarious. The effects are more prominent with the more short-lived vectors so that a rise in temperature could increase the relative importance of minor vectors. For example, if *P. falciparum* is being transmitted by two vectors, one long-lived ($p = 0.95$) and responsible for 90% of transmission and the other short-lived ($p = 0.8$) and responsible only for the remaining 10% of transmission, a 2 °C rise in temperature from 22 °C would double the relative importance of the minor vector, in this example, whereas if the initial temperature were 20 °C, a two degree rise could increase the share of transmission by the 'minor' vector to 35%.

Of course, temperature does not affect one aspect of malaria transmission in isolation. Rising temperatures change vector behaviour. In general, feeding will become more frequent, and because transmission is proportional to the square of the biting frequency this will affect R_o substantially. At the higher temperature the frequency of oviposition will increase and the duration of the larval stages will be shortened, so allowing mosquito populations to build up more rapidly. However, survival as an adult will tend to fall and life expectancy to be reduced, unless there is a corresponding increase in relative humidity. This effect on survival will reduce transmission, and has the capacity to do so markedly, because transmission is proportional to the nth power of the chance of survival through a day where n rarely falls below 10. The upshot of this messy complexity in most situations is that transmission will increase as temperature rises, but not to a precisely predictable level, and that the effect will be greatest at the temperature limits of transmission, in the absence of deliberate public health action.

Precipitation is predicted by the IPCC to rise by about 5% on aggregate, but with marked local variation. This will also tend to increase the transmission of malaria, in two ways. Higher relative humidity prolongs anopheline life (increases p), which has a marked effect on R_o, but the increased rainfall may also increase the number of mosquito breeding sites or their persistence for a greater part of the year, so increasing mosquito density. Transmission is directly proportional to density, so this will have an effect, but if the increased precipitation also makes part of the Sahel habitable for mosquitoes for more of the year than before, the transmission effect would be much greater. *Anopheles culicifacies*, a major vector in Sri Lanka and parts of India, breeds mainly in ponds formed from drying-up streams, and the severe epidemic

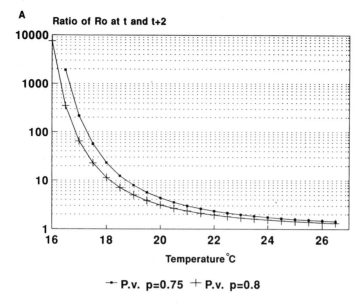

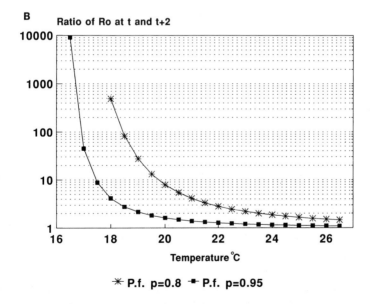

FIG. 4. The proportionate increase in R_o, the basic case reproduction rate of malaria, caused by a 2 °C rise in temperature reducing the duration of the extrinsic cycle (n); other variables are assumed to be unaffected. Part A shows the changes for *P. vivax* (P.v.) at $p = 0.75$ and $p = 0.8$, and Part B shows those for *P. falciparum* (P.f.) at $p = 0.8$ and $p = 0.95$.

falciparum malaria of Sri Lanka tends to occur in drought years. The relation to climate is complex, though relatively well studied.

Biological factors are not the only determinants of malaria transmission, however, and it is highly unlikely that the potential increases in transmission resulting from global warming would be realized in full, because malaria does not currently extend to its climatic limits but is limited by deliberate human action. Europe is free of indigenous malaria (there is some *P. vivax* in the Asian part of Turkey) as a consequence of massive eradication programmes carried out in the years following 1945. Prior to then most of southern Europe was highly malarious in summer, especially parts of Italy, Greece, Albania and the islands of the Mediterranean. Indeed, malaria was endemic in The Netherlands, but here there was indoor transmission, with the vectors living around the stoves used to heat the houses, such that transmission occurred particularly in the cooler months.

In summary, the main effects of global warming on the transmission of malaria are likely to be:

(i) Transmission at higher altitudes in malarious developing countries such as Ethiopia, Rwanda and Burundi, and possibly Kenya, where transmission will spread up the mountains.

(ii) Spread of transmission towards temperate regions in areas of low economic development where malarious and non-malarious areas are adjacent and control measures are absent.

(iii) Increasing importance of currently minor vectors in areas of limited transmission.

(iv) Increased frequency of small outbreaks of introduced malaria in countries where it is currently not endemic.

Vector-borne diseases imported from the tropics to Europe

The preceding discussion focuses on the gradual extension of the limits of transmission. What are the specific hazards of imported malaria for Europe and the likelihood of localized outbreaks secondary to imported malaria? Over 10 000 cases of malaria are probably imported annually into Europe. We know that at least 2332 cases were imported into the UK in 1991 *and* reported, and that 1314 of these were of the potentially lethal falciparum malaria. Future trends depend on four factors: the level of overseas travel, the level of malaria transmission in countries visited, the amount of resistance of the malaria parasites to the prophylactic drugs used, and the level of compliance by travellers with advice on how to avoid being bitten by infective mosquitoes and regular use of chemoprophylaxis. Drug resistance will go on spreading, control programmes are not currently very effective, and compliance improves only slowly. The extent of overseas travel is more variable in the short term, and

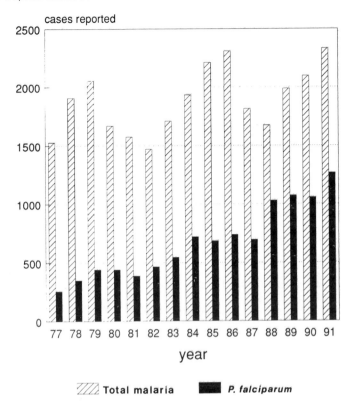

FIG. 5. Incidence of malaria imported into the UK and reported to the Malaria Reference Laboratory, 1977–1991.

economic recession is good for limiting imported malaria! But the data for *P. falciparum* show an inexorable rise since at least 1977 (Fig. 5) and this trend is likely to continue.

Local spread from imported malaria follows one of two different patterns. *P. vivax* is more likely than *P. falciparum* to produce secondary cases because the extrinsic cycle requires a minimum temperature of only 14.5 °C, cooler than for *P. falciparum*. There have been no such secondary cases for 30 years in the UK although appropriate anophelines are found in East Anglia and elsewhere. The risk is much greater in southern Europe and especially the Mediterranean littoral. There has, however, been an outbreak in the UK of what is known now as 'airport malaria', and many more such outbreaks have been reported around continental international airports. Here, the infected immigrant is a stowaway mosquito and it is interesting that the UK outbreak occurred during one of the hottest summers on record. It is believed (Whitfield et al 1984) that an *Anopheles gambiae* already infected with *P. falciparum* travelled from

West Africa, escaped at Gatwick airport, and bit two people who did not leave the UK. They were both severely ill. The anopheline did not breed (presumably) but it needed a climate warm enough to survive and feed. Such incidents are likely to increase.

In summary, the tropical environment is undergoing rapid change, with consequences for health both locally and on a wider scale. Climatic change may increase the geographical extent of vector-borne diseases, while the many other environmental changes are leading to complex and locally variable responses on the part of pathogens, vectors, and their human hosts.

Acknowledgement

I am grateful to Professor R. M. May and the Oxford University Zoology Department for the sabbatical environment in which this article was written. It forms part of the ODA Programme in Tropical Diseases Control.

References

Andrewartha HG, Birch LC 1954 The distribution and abundance of animals. University of Chicago Press, Chicago, IL

Bradley DJ 1991 Malaria. In: Feachem RG, Jamison DT (eds) Disease and mortality in sub-Saharan Africa. Oxford University Press, New York, p 190–202

Bradley DJ 1993 Environmental and health problems of developing countries. In: Environmental change and human health. Wiley, Chichester (Ciba Found Symp 175) p 234–246

Houghton JT, Jenkins GJ, Ephraums JJ 1990 Climate change, the IPCC scientific assessment. Cambridge University Press, Cambridge

Macdonald G 1957 The epidemiology and control of malaria. Oxford University Press, London

McGregor IA 1991 Morbidity and mortality at Keneba, The Gambia, 1950–75. In: Feachem RG, Jamison DT (eds) Disease and mortality in sub-Saharan Africa. Oxford University Press, New York, p 306–324

Molineaux LH 1990 Epidemiology of malaria. In: Wernsdorfer WH, McGregor IA (eds) Malaria. Churchill Livingstone, London, p 913–998

UNICEF 1985 Analysis of the situation of children and women. UNICEF, Dar es Salaam

Whitfield D, Curtis CF, White GB, Targett GAT, Warhurst DC, Bradley DJ 1984 Two cases of falciparum malaria acquired in Britain. Br Med J 289:1607–1609

World Bank 1992 World development report, 1992: development and the environment. Oxford University Press, New York

DISCUSSION

Sutherst: What will be the situation with regard to malaria in 30 years?

Bradley: That will depend on whether we have a vaccine. It is clear that the rate at which parasites are becoming resistant to the drugs that are available and to the new drugs is exceeding the rate at which new drugs are being introduced. Moreover, the drugs that are being introduced are expensive, for good reasons sometimes. Bed nets impregnated with pyrethroids are likely to be effective for a period, but culicines are already beginning to generate resistance to pyrethroids, and no doubt the anophelines will follow shortly. Historically, by and large, pathogens have not been good at generating resistance to vaccines, but we don't have a vaccine. The enthusiasm for a vaccine and the large amounts of research that were done meant that a lot of promising candidate vaccines were rushed through, perhaps in too much of a hurry, and proved ineffective in people. The likelihood of an effective vaccine appearing is quite high, but we won't know what it is until it does! Once we have that breakthrough, we can devise ways to produce the vaccine economically. Until then, we are going to be on a losing wicket.

Klein: I am a little reluctant to believe that the economic problems of vaccine production will be solved. As we have seen before, there are many basic ideas, but the vaccines are required in developing countries where people are unable to pay for their application. Therefore, international organizations such as WHO have to cover the expenses. The knowledge and interest in these scientific issues in the developed countries is high, but as long as we haven't solved the problem of who produces and applies the vaccine and who pays, there can be no economic solution.

Bradley: I think the market for a malaria vaccine in developed countries is sufficient to provide some degree of encouragement. Once the researchers funded by WHO and national research councils reach the point of having something promising, I think it would be taken up. History tends to show that once something does get taken up, there are ways of pushing the price down. That certainly happened with the anti-schistosomal drugs. Ivermectin is free to people in the Third World because the manufacturer can make enough money selling it for use in cattle in developed countries. First we need to solve the technical problem, the scientific problem, then there are ways to force the price down.

Edwards: Dr Hill's recent work has suggested that the best prospect for a vaccine is a short peptide. A short peptide can be produced biochemically very cheaply, and it may even be possible to take it by mouth or as snuff. This would be a cheap form of vaccination.

Bradley: It would be cheap if it worked and only one or a few peptides were needed. Patarroyo has a vaccine of this type that appears promising.

Edwards: Even a cocktail wouldn't cost much more because the main cost is in delivery rather than production. Ten amino acids can be strung together for a few cents per amino acid; this is a trivial problem for the chemical industry.

Elliott: Dr Bradley, you showed that the lifespan of the mosquito is a crucial element in the transmission equation. Is there any way in which different mosquitoes with shorter lifespans could be introduced, perhaps by natural means, such as through climatic change, or by intervention?

Bradley: There isn't any subtle way of introducing mosquitoes with shorter lifespans. One can wipe out the habitats of certain vectors, but unfortunately *Anopheles gambiae* is a puddle breeder, and eradicating puddles from Africa is not a terribly practical proposition. It may be possible with other vectors, particularly in Asia. The tea-growing industry in India and Bangladesh is dependent on environmental control of the particular vectors in the area by what's called species sanitation—the place is made unsatisfactory for breeding by the local, fussy, mosquito vector species.

There is hope in looking for genes which could make the mosquito resistant to malaria. It's not particularly in a mosquito's interests to be susceptible to malaria. It doesn't actually harm it, it doesn't shorten its life appreciably, but it doesn't benefit the mosquito either. This approach was tried some 10–15 years ago, but it was far too early, because the rest of science wasn't up to it. Now one could, in the experimental malarias, find the genes, then find analogous genes or transplant resistance genes using infectious plasmids. Louis Miller at NIH in the USA and various other groups are pursuing this approach.

Lake: Are there any ethnic, physiological or immunological differences in susceptibility to malaria between different sectors of the population?

Bradley: P. vivax, by and large, will not infect the indigenous people of West Africa. This is not a geographical difference; in the USA, in areas which were highly malarious before eradication a few decades ago, the white people got vivax malaria, and the black people who originated in West Africa did not. The Duffy blood group receptor acts as a sort of receptor for *P. vivax*. In West Africa, the surface of the red cell is Duffy-negative, so *P. ovale* has probably evolved to fill that niche.

There are some people with extremely peculiar red cell antigens, less than one in a million in Britain, who have difficulty being infected with *P. falciparum*. For reasons which are unclear but must be sufficiently fundamental to the structure of the red cell, that characteristic has not spread through populations where it would clearly be beneficial.

The sickle cell gene is the classic example of a resistance gene to malaria. A single genetic change reduces the mortality from falciparum malaria. Because the homozygous state, possession of two sickle cell genes, is lethal, the advantage offered with regard to malaria must be great. In some parts of Africa, 10–20% of the population carry the sickle cell gene. To maintain such a frequency of the sickle cell gene, about 10% of the children without the sickling gene must have been dying of malaria. These are genetic changes which provide protection. The only explanation that has yet been produced for the existence of some of the haemoglobinopathies is that they seem to provide protection against malaria. The remarkable thing is that, although malaria parasites have been around for several million years, humans have not evolved more complete resistance mechanisms. The malaria parasites must be able to evolve quickly enough to keep ahead.

Lake: Does nutritional status affect susceptibility to malaria?

Hautvast: Anaemia provides protection against malaria. When you treat anaemic people with iron, the prevalence of malaria increases. There is increasing evidence that certain vitamins, vitamin A in particular, may play a role with regard to the severity of the disease by affecting production of cytokines.

Bradley: At the level of specific nutrients the situation is complicated, and is just being sorted out. The striking thing is that malaria epidemics come at the end of the famine, at the re-feeding stage.

Rabbinge: Will the danger of epidemics increase considerably with environmental change? It might increase when temperature increases because the intrinsic rate of increase of the mosquito population is likely to be greater. However, the intrinsic rate of increase may be affected by natural enemies, which may be affected in the same way as or more or less than the malaria parasites.

Bradley: That was why I stressed that the effects of climate change, other than on the extrinsic cycle duration where we can be precise, are extremely uncertain. At the edge of its range, malaria is, by and large, not determined by climate. There are good reasons why there is no malaria in Europe. Deliberate efforts were made to get rid of it, and would be made again if it came back.

Rabbinge: In The Netherlands, we had a malaria epidemic at the time when the Zuider Zee was closed and became the IJsselmeer in the early 1930s. The salt water of the sea became brackish, which enabled the mosquitoes which carried the malaria parasites to survive. The malaria epidemic nearly became a pandemic and much research on malaria began in The Netherlands.

Bradley: At the edges of its range, other reasons will prevent there being great malaria outbreaks. Where environmental change is more important is in Southeast Asia and Brazil, where forests are being cut down, and malaria is occurring at the edge of the forests. In irrigation schemes in Asia, but less so in Africa, apart from in Sudan, the water resource developments will change transmission.

In South-east Asia, a whole series of environmental factors come together to increase forest fringe malaria—human activity at the forest edge, rapid disturbance of the forest, lots of illegal activities, because the forests are mostly near national boundaries, a lot of smuggling, a lot of illegal gem mining, a lot of logging. These activities create breeding places; the holes dug for tin and for precious stones fill up with water and the mosquitoes breed in them. These areas are also politically unstable, populated by ethnic minorities, and the government's malaria control programmes can't get in because of security risks. This complex of environmental change and socio-political change leads to great outbreaks of malaria.

The situation is similar in Brazil, though many sites are less complex. Gustavo Bretas is working on the determinants of malaria in the Brazilian rain forest fringe, a task facilitated by satellite and radar images of the area, made easy to obtain by the Brazilian government's purchase of their copyright. Survey and control data for malaria are superimposed on topographical and vegetation images. The interesting finding so far has been that the level of malaria does not depend only on proximity to the forest fringe; the highest rates are found in low-lying parts of the forest edge. The combination of being near the trees and in a low-lying area subject to back-flooding makes for successful populations of *Anopheles darlingi* that transmit malaria. If you got rid of the forest totally, there would be no malaria, but I don't feel that would be a desirable approach.

Mansfield: Do the models used to predict shifts of insect populations take photoperiodism into account? The life cycles of many insects are determined photoperiodically. Surely the fact that temperature *will* change but photoperiodism *won't* at a particular latitude has to be taken into account. It may not be so easy without some evolutionary change in an insect population for it to move along a temperature gradient.

Sutherst: That's an important point. Photoperiodism is one of the most plastic of the attributes of insects. It seems to be easy for them to adapt to new day-length regimens. The response of ticks, for example, to day length at different latitudes differs within very short distances. The effect is polygenic with frequent recombinations which permit adaptation. If there were a very strict photoperiod and little variability, a population could be lost in a year with an early or late winter. The large amount of variation means that a proportion will always survive through to the next year. I would think there would be enough genetic variability for easy adaptation to change.

Rabbinge: Photoperiodicity is an important subject, especially in plants. Until about 30 years ago we had virtually no corn at this latitude, mainly because the plants reacted so strictly to day length. The breeders were aiming to breed varieties less sensitive to temperature, but in fact they were selecting for day-neutral varieties which can grow at many latitudes.

Insects have plasticity, and there are geographical races whose reactions are photoperiodically determined. Different geographical races of the Colorado beetle, for example, react differently to day length.

What is the situation in the mosquitoes which carry malaria? I believe the mosquito is a typical example of a day-neutral insect.

Bradley: I don't know about those at the edge of their range, but that is the case with those living in the true tropics.

Zehnder: You said that malaria rises after famine when people start to eat normally again. They are in a sort of intermediate stage in which they are strong enough to survive but there is still a hangover from the time of the famine. We have said that the important thing about soil pollution is not what's in the soil but what finally comes out of the soil into the ground-water or into plants. Would a low level of toxicity maintained for quite a while affect people's susceptibility to these kinds of diseases?

Bradley: Very little is known about this because studies are usually done in experimental systems looking at fairly gross changes over a short time. Chronic experiments are rare.

Zehnder: There are situations where there is natural pollution, as a result of geological features, such as those areas in the USA where selenium and arsenic levels are high.

Bradley: With most infections there are other variables which are large in relation to trace metals. How do you control for these other effects? You are looking for a small effect among the massive local effects resulting from differences in human behaviour. This problem is similar to that of looking for genetic variations in susceptibility to infection, which proved intractable until very recently, simply because there was so much 'noise' from environmental factors that have a direct effect. The effects of trace elements are much more easily demonstrated with non-communicable diseases.

Hoffmann: Are there any clear-cut differences between the rates of malaria transmission in urban and rural areas?

Bradley: That depends on location. There are enormous local variations, and this is true of vector-borne diseases in general. For example, in Africa in towns there tends to be dirty water rather than clean water, and therefore there are less anophelines and consequently less malaria transmission than in the countryside. In contrast, in India, there is an urban mosquito, *Anopheles stephensi*, which is an extremely good vector. Transmission is highest at the forest edges and fairly low in the plains, but is almost as high as in the forest edges in a highly urbanized area such as Surat because of *Anopheles stephensi*. With most parasites, the local variations arising from local features will overshadow any generalizations.

Zehnder: I appreciate what you are saying, but there are people here who actually have the data. Professor James could produce a map of nutritional status, Professor de Haan has information on soil pollution, and you have shown maps of the distribution of certain diseases. Has anybody tried putting these maps on top of each other?

Bradley: This has been done to some extent, but the results have usually been unhelpful, because local changes are ignored. This approach has a bad track record. Take schistosomiasis as an example. Joel Cohen believed he could show that one species of schistosome protected against infection by another, by analysing data from an area where there were people with one sort, people with the other sort, and people with mixed infections. There were far less people with mixed infections than expected, so he had inferred cross-protection. In fact, the area was not homogeneous; the people who lived near the river got one sort, and the people who lived near ponds inland got another sort. The explanation was an ecological one.

James: The other problem is that from a nutritional and a toxicological point of view the instruments currently used to specify the state of individuals or populations are incredibly crude. In the nutritional world one doesn't like to say it, but the truth is that one is still using concepts and indices which were developed in the 1950s. Sophisticated, sensitive tests which show dose–response effects have really not been applied in the nutritional world at a population level, for example in the Third World countries. The epidemiologists wanting to look at the effects of the physiology and nutrition of individuals end up in despair because the scientists don't know what to do or what measurement to provide them. The epidemiologists therefore develop cruder and cruder instruments, in an attempt to get some way of looking at nutritional or health status.

Sutherst: With livestock there are clear relationships between the level of nutrition and resistance to parasites. I would assume that the same would apply in general to humans. It's more difficult to talk about specific soil pollutants than about protein deficiency, for example. I would have thought that a problem of contaminants affecting resistance would show up only if the contaminant were having some other clinical effects.

Zehnder: Perhaps my question wasn't quite clear. If you put one sort of map of Africa over another sort of map of Africa, the result would be unhelpful. If a more regional approach were taken, with more detailed information about ecological and structural parameters, for example, the useless information could be drastically reduced. For example, it is not only the amount of food which is important, but also the quality of the food. If the soil is polluted there is a substantial chance that the pollutants will enter the food. However, the amount entering food depends on the buffering capacity of the soil, which in turn is a function of the soil's composition. People taking in toxins over a long period of time cannot feel as healthy as they would do if they were not taking in toxins.

Elliott: I agree that we should maintain the buffering capacity of the soil— it's sound ecological practice—but there is no evidence to support your assertion with regard to human health. The evidence we have is that during the period over which there has been massive introduction of chemicals, life expectancy has increased in industrial nations. The overriding health problems now are more to do with over-eating and over-consumption than with industrial pollution, though that doesn't mean that we shouldn't be cautious.

Klein: The real problem we face is getting good, representative data about health. Because of data protection legislation and other regulatory restrictions in Germany we are unable to develop even a good cancer register. This is true both nationally and regionally. While we agree in principle that we should move towards regional health surveys, there are tremendous problems and restrictions in doing so.

Sutherst: We have used the CLIMEX model successfully to identify places in South America and Africa, for example, where there are large populations of ticks on cattle. This was done simply by looking at areas that have poor nutritional conditions, because we know that nutritional stress in cattle is a primary factor determining the size of tick populations. The difficulty with such an approach with humans is that we store and transport food and so override the natural seasonal shortages.

James: We also need to recognize that if we are going to start dissecting out these issues we are going to have to take an approach somewhat different from that of the toxicologist studying acute effects. The problem of osteoporosis, for example, the thinning of bones, which occurs particularly in elderly women, may actually be substantially affected by the long-term body burden of lead, which we know has a fundamental effect on key enzyme processes involved in the development of the bone matrix. The thinning of elderly bones is not, as everybody had assumed, necessarily a feature of calcium metabolism; the protein matrix is in fact involved.

Klein: Why then is osteoporosis more common in females than in males?

James: This is because they get to a critical low level of bone thickness earlier in old age than men, having had in their youth a much smaller total bone mass than males. Crumbling and microfracture of the bone matrix begins when there is a limited amount of matrix that can't withstand the stresses. The best way to avoid osteoporosis is to be obese!

There are also more subtle effects. Alcohol, for example, actually changes the way in which toxicants are processed in the liver. There are also extraordinary interactions, which are only now being looked at, between the effects of drugs and alcohol. Intake of alcohol entrains a completely different metabolic system, which modifies the rate at which drugs are used, and nutrition interacts there too. The standard therapeutic approaches of doctors in the West have to be seen in a different context when dealing with people who are relatively undernourished. Trying to incorporate the issues of heavy metal pollution into that complex of actions is a nightmare. We can accept that there are interactions, for example, between cadmium, zinc and copper metabolism which are now being looked at. Zinc is crucial in the replication of cells and in immune functioning, yet the interactions between zinc and prevailing levels of cadmium or copper have not been looked at in a coherent way. Some ideas can be derived from acute poisoning episodes, but people haven't applied modern understanding of immune function and infection in this new context. That's something which is quite hard to do, but will be necessary if we are going to answer the real challenges.

Zehnder: It seems that the tropics are the hot spots for pollution and will also be most affected by changes in temperature and pests.

Avnimelech: The hot spot of pollution is northern Europe.

Zehnder: I am not so sure about that. Pollution is closely monitored in this part of the world, whereas it is not in others. Absence of data does not mean there is an absence of pollution. Is it wise to put our main emphasis on monitoring our own environment, or ought we to identify the most acute environmental problems and give them first priority, regardless of the geographical area where they occur?

Bradley: Somewhere where there's little research investigating both infections and chemical pollution, and where it might be productive, is industrializing India, where there are high rates of infections still coupled with high levels of heavy metals and industrial pollution. As far as I know, no one has systematically looked at that.

Edwards: We have discussed heavy metals which are not necessary and are possibly detrimental, and also metals such as copper, iron and zinc, which are actually essential elements. One well-understood mechanism of protecting against parasites in the blood is to withhold iron from them. Haptoglobins and transferrins bind iron strongly and make iron unavailable to parasites. It's quite likely that there are similar mechanisms to deprive them of copper, zinc and other substances which are essential to their well-being. This could be one reason why when people are well fed after deprivation they are at greater risk from parasites. Parasites usually prefer healthy people. Excess copper or zinc in the diet would probably upset the individual's mechanism for withholding these necessary trace elements from invaders.

de Haan: Do you think there would be competition?

Edwards: Too much iron is certainly detrimental. Everybody thinks iron is good for you, and it's doled out to pregnant women. The general view is that being below the average in haemoglobin is bad for you. This is probably true if you live in The Netherlands, but it's not at all obvious that the same is true in India, for example, where a person might be much better off not having any excess of iron in the blood available to plasmodia and other parasites.

de Haan: In The Netherlands we suddenly found a specific location where food quality standards with respect to cadmium could not be met. We found out that the cadmium content of the soil in this area itself was normal, in fact lower than in some other ares where there were no problems meeting the food quality standards. Where there is cadmium pollution there is always zinc pollution at a much higher level. We finally realized that the problem in this area was that there was pollution by cadmium only, without zinc. The combination of pollutants is important.

Hoffmann: I am curious about the effect of air-borne environmental contaminants on mosquitoes, for example. We are always concerned about the effect of ozone on humans and animals and plants. Have there been any systematic studies of the effects of such ambient chemical toxicants on insect disease vectors?

Sutherst: The only studies I am aware of are on the effects of water waste containing detergents. *Culex quinquifasciatus* is more tolerant of this than other mosquitoes, which helps it to take over and transmit filariasis (Service 1989). I don't know anything about the effects of aerial pollutants.

Mansfield: Insects are relatively resistant to air pollutants. Aphids, for example, are more affected by the action of air pollution on the plants they feed on, than they are by the air pollutants themselves.

Reference

Service MH (ed) 1989 Demography and vector-borne diseases. CRC Press, Boca Raton, FL

Human respiratory disease: environmental carcinogens and lung cancer risk

J. C. S. Kleinjans, J. M. S. van Maanen and F. J. van Schooten

Department of Health Risk Analysis and Toxicology, University of Limburg, PO Box 616, 6200 MD Maastricht, The Netherlands

Abstract. Inhalatory intake of environmental agents may have adverse effects on health, the lung being the first target. Therefore, an increased risk of lung cancer and respiratory disease is in general considered as an indication of environmental health problems related to exposure to industrial emissions, traffic exhaust and smog. Classical epidemiological studies of the association between exposure to ambient air pollutants and respiratory dysfunctions and studies with laboratory animals have failed to demonstrate the distinct proof of risk for the general population that would be needed to form a basis for high impact environmental policy measures. Here, as an example, we describe the uncertainty in assessing risks of lung cancer associated with environmental exposure to polycyclic aromatic hydrocarbons. The recently introduced methodology of molecular cancer epidemiology is considered to yield more information on the relationship between exposure to environmental carcinogens and tumour development. Recent advances in the study of carcinogen (polycyclic aromatic hydrocarbon) dosimetry at the DNA level in combination with proto-oncogenic activation in humans are described.

1993 Environmental change and human health. Wiley, Chichester (Ciba Foundation Symposium 175) p 171–181

An introduction to the field of molecular epidemiology

The respiratory tract is a major port d'entrée for environmental agents which may have adverse effects on health, the lung obviously representing the primary target organ. Therefore, an increased risk of lung cancer and chronic respiratory disease is in general considered to be an indication of environmental health problems related to exposure to industrial emissions, traffic exhaust, smog or contamination of indoor air, by radon, for example. These risks are regarded with much concern. One of the important goals within the WHO's regional strategy for Health for All is that by 1995 all people of the European region should be effectively protected against recognized health risks from air pollution

(WHO 1987). Execution of this particular strategy, however, would require enormous economic investment, and policy decision still suffers from lack of sufficient scientific knowledge on the causal relationship between the inhalatory intake of environmental pollutants and the development of lung disease. For example, epidemiological studies on exposure to smog and the incidence of acute or chronic respiratory dysfunctions fail to demonstrate the distinct causality that would be needed as a basis for high impact policy measures designed to reduce road traffic. Environmental health research should therefore be directed towards uncertainty reduction. A major benefit is expected from the introduction into epidemiological studies of new methodology in which biological markers are used to assess the actual dose of a chemical to which tissues are exposed, and early biological effects indicative of a risk of pathological developments in the longer term; this approach differs from that of classical epidemiology, in which data on air quality are linked with disease incidences in the general population. So far, experience in the application of biomarker analysis to epidemiological issues, an approach generally but erroneously referred to as molecular epidemiology, has been predominantly gathered in the field of chemical carcinogenesis.

Chemical or physical induction of damage to DNA by environmental factors present at target sites is considered to be a first step in the process of environmental carcinogenesis. First generation biomarker analysis has therefore been focused on the detection and measurement of irreversible genetic errors such as point mutations, chromosomal aberrations and sister chromatid exchanges, mostly in cases of occupational exposure to known or suspected carcinogens (Hulka et al 1990) but also in general subpopulations at increased risk of exposure (Kleinjans et al 1991). As regards selectivity, these cytogenetic endpoints have been acknowledged to be of limited relevance: they are affected by a broad spectrum of environmental genotoxic agents and can be influenced by person-bound factors such as age and sex; furthermore, inter- and intra-individual variability can be considerable. However, careful design of the intended biomarker study, as is common in epidemiological practice, may overcome most of these difficulties. Obviously, analysis of this type of biomarker does not provide information on the actual dose of the carcinogen reaching the tissues. Furthermore, sensitivity of these markers is generally overrated: on reviewing the available literature, one notes that in active smokers the lowest effective dose of cigarette smoke is quite similar whether peripheral lymphocyte chromosome aberrations or lung cancer incidence is used as the outcome measure. The same holds true for exposure to vinylchloride in occupational settings: the occurrence of chromosome aberrations and the incidence of liver cancer indicate a similar lowest effective dose. A major advantage of genetic biomarker analysis is that statistically reliable data can be obtained from relatively small but properly selected sample sizes. Recent advances in our knowledge of the mechanisms of chemical carcinogenesis, in combination with the development of new methods for endogenous dosimetry, seem to be yielding a second generation of genetic biomarkers which may more usefully be applied to the investigation of environmental cancer risk.

Polycyclic aromatic hydrocarbons and lung cancer risk

Polycyclic aromatic hydrocarbons (PAHs) are formed mainly as a result of pyrolytic processes, particularly the incomplete combustion of organic materials, as well as during carbonization. As a class of potentially carcinogenic agents occurring ubiquitously and in varying concentrations in the environment, they have gained much attention from toxicologists and epidemiologists, and also from environmental policy-makers. About 500 different PAHs have been demonstrated to be present in city air, but most measurements have been made on benzo(a)pyrene (BaP) because of its probable carcinogenicity in humans. Annually averaged urban air concentrations of BaP in Western countries have been shown to vary between 1 and 5 ng/m³; the estimated inhalatory intake of BaP is 40 ng per day (this assumes a pulmonary deposition rate of 40% from 20 m³ air inhaled per day by a person outdoors all day) (WHO 1987). An individual's lifetime risk for developing respiratory tract cancer through inhalation of BaP has recently been calculated to be 3.3×10^{-3} per μg BaP/m³ (Collins et al 1991). This suggests there would be about 16 respiratory tract cancers in a population of 10^6 persons exposed to 5 ng BaP/m³ for their lifetime. Additionally, intake of food contributes significantly to the body's PAH content; estimated intake of BaP from food per person ranges from 0.29 to 1.3 mg per year in European countries (Preussmann 1985). On the basis of current cancer risk assessment models for oral intake of BaP (Slooff et al 1989), the estimated incidence of cancer is 3–26 in a population of 10^7 persons consuming BaP at this level throughout their lives. The environmental occurrence of PAH typical in Western countries therefore imposes a considerable cancer risk for the general population.

These estimates of cancer rates are predominantly based on animal experiments, so include uncertainties regarding species-to-species and high dose-to-low dose extrapolation. Risk assessment based on epidemiological observations on gas workers and coke oven workers is also not without problems because such people were simultaneously exposed to PAHs other than BaP, as well as to other carcinogens; furthermore, actual levels of exposure to BaP could not be determined in these studies (Collins et al 1991). In view of its presumed carcinogenicity no safe level of BaP exposure can be recommended; the risk of BaP-induced cancer remains while emission via diesel exhaust and cigarette smoke still continues (WHO 1987).

Measurement of exposure to PAH at the DNA level

It is believed that the use of molecular and biological markers may be helpful in addressing some of the major problems in environmental health risk assessment.

The uncertainties about analysis of PAH-related environmental carcinogenesis risks described above indicate that knowledge of the endogenous level of exposure to, specifically, BaP is necessary. This could be achieved by measuring levels of PAH–DNA adducts, in, for example, DNA derived from peripheral lymphocytes obtained by venous puncture from individuals exposed to PAHs. Adducts are fairly stable complexes of xenobiotic chemicals covalently bound to cellular macromolecules such as DNA, RNA and proteins. In general, adduct-forming compounds, or their reactive metabolites, are electrophilic molecules which can easily react with nucleophilic DNA sites. In order to become electrophilic, BaP has to be transformed, through a reaction with molecular oxygen catalysed by intracellular cytochrome P-450 enzyme complexes, to its major reactive metabolite 7,8-dihydroxy-9,10-epoxy-7,8,9,10-tetrahydrobenzo-(a)pyrene (BPDE), which may subsequently bind covalently to the DNA nucleoside guanosine. Therefore, determination of the level of DNA adduction provides information on the integrated result of exposure, absorption, metabolism and excretion, the kinetics of DNA adduct formation and repair of the introduced DNA lesions. A variety of sensitive and relatively selective assays are available to identify BPDE–DNA and other related PAH–DNA adducts, including ^{32}P-postlabelling and immunochemical techniques (Van Schooten et al 1990). Immunocytochemical methods may be used to detect BPDE–DNA levels as low as one in 10^8 nucleotides, and in ^{32}P-postlabelling assays a detection limit of one adduct in 10^9–10^{10} nucleotides has been realized. Adduct levels have been determined in relation to tobacco consumption, in people who work in coke-ovens and foundries, and in relation to urban versus rural residence (Goldring & Lucier 1990, Marx 1991, Perera et al 1991, Van Schooten et al 1990). Because respiratory tract tissues, i.e., the tissues damaged by these particular chemical carcinogens, are not readily accessible for DNA adduct studies in humans, white blood cell (WBC) DNA has commonly been used instead. In patients with lung cancer, levels of PAH–DNA adducts in pulmonary tissues correlate with cigarette consumption, but no relationship between adduct levels in WBCs and lung tissues has been found (Van Schooten et al 1992). Overall, no consistent quantitative relationship between cigarette smoking and lymphocyte PAH–DNA levels has been found, but in iron foundry and coke oven workers increased WBC PAH–DNA adduct levels have been found repeatedly. Furthermore, oral and dermal intake of PAH may lead to formation of PAH adducts in WBC DNA (Rothman et al 1990, Paleologo et al 1992). The route of exposure seems to influence the formation of WBC PAH–DNA adducts profoundly, which emphasizes the limitations of the use of lymphocytic DNA for monitoring human exposure to inhaled carcinogens; further research in this area is required.

Although this field is promising, there are several problems. Generally, PAH–DNA adduct levels in target tissues are related to the ultimate carcinogenic response. However, there is no simple quantitative relationship between DNA adduct formation and tumour induction in human or rodent tissues; similar adduct profiles can be observed in susceptible and resistant tissues. Differences in the persistence of the adducts may be involved, but it is unclear whether persistence is related to susceptibility or target tissue specificity. Moreover, the relationship between adduct measurements in a surrogate tissue such as the WBC and the potential risk at the target site—the lung—has to be determined. Studying the dose–response relationships between PAH–DNA adducts (in target and surrogate tissues) and tumorigenic events may increase our knowledge of the association between exposure to environmental carcinogens and lung cancer risk considerably.

PAH-induced proto-oncogene activation

The ultimate goal of molecular epidemiology is to assess cancer risks at the individual level. This is in contrast to classical epidemiological methods, although these have been very helpful in identifying groups of persons at increased risk of developing cancer (Marx 1991). Carcinogen dosimetry alone will not be enough. All DNA adducts are potential markers for exposure of a target site, but not all DNA adducts will be indicative of adverse effects on health such as cancer (Bond et al 1992). As already mentioned, carcinogen dosimetry at the level of DNA may therefore be linked with the analysis of parameters indicating the onset and development of cancer. Early techniques for measuring presumably precarcinogenic events, such as the detection of chromosomal aberrations, micronuclei and sister chromatid exchanges, have shown increases in levels of genetic damage in peripheral lymphocytes from carcinogen-exposed subjects, in occupational settings for example (Ashby 1988, Hulka et al 1990). However, carcinogen-induced chromosomal aberrations or sister chromatid exchanges *in vitro* did not appear to correlate significantly with the formation of tumours induced by the same carcinogens in rodents (Rosenkranz et al 1990). Furthermore, cytogenetic analyses carried out in the course of a prospective study on the effects of occupational exposure on cancer revealed only a slight, statistically significant association between the occurrence of cancer and the extent of peripheral lymphocyte chromosomal aberrations (Sorsa et al 1990). Also, Perera et al (1989) measured PAH–DNA adducts and sister chromatid exchanges in peripheral lymphocytes from patients with lung cancer and controls, and found increased PAH–DNA levels in currently smoking lung cancer patients, but no correlation with peripheral lymphocyte sister chromatid exchanges. Overall, a clear dose–response relationship between PAH–DNA adduct levels and these cytogenetic parameters has not yet been described. The relevance of analysis of these types of genetic damage to the prediction of cancerous events to come therefore seems limited.

Over recent years, fundamental cancer research has provided much information about activation of proto-oncogenes as an early process in carcinogenesis.

Proto-oncogenes encode proteins that seem to have a pivotal role in cell growth and differentiation; DNA damage leading to proto-oncogenic amplification or mutation may raise intracellular levels of these proto-oncogene proteins or alter their structure, thereby interfering with cell growth-regulating pathways and ultimately inducing malignant cellular transformation and neoplasia. Oncogene activation by chemical carcinogens has been demonstrated in several laboratory animal models (Balmain & Brown 1988). Furthermore, newly developed molecular biology methods such as the polymerase chain reaction technique appear to be highly sensitive and can be used with relatively minor amounts of DNA. The analysis of measures of oncogene activation in carcinogen-exposed humans may provide a sensitive marker which is also highly relevant because of its mechanistic role in tumorigenesis.

BPDE appears to form adducts with human Ha-*ras* proto-oncogenes *in vitro* (Dittrich & Krugh 1991) and our own studies demonstrate amplification of Ki-*ras* and over-expression of *myc* in human fibroblasts exposed to BaP. BaP-induced skin and liver carcinomas in laboratory rodents show a high frequency of *ras* gene point mutations (Balmain & Brown 1988). Furthermore, mutations in the *ras* gene were more frequently found in lung adenocarcinomas from smokers than in those from non-smokers (Slebos et al 1991). Perera et al (1991) showed that a group of patients with lung cancer, all current or former smokers, had increased serum levels of one or several proteins encoded by oncogenes, suggestive of proto-oncogenic amplification or over-expression; those patients still smoking also had increased levels of PAH–DNA adducts (but there was no significant correlation between DNA adduct levels and serum oncoprotein concentrations). This study additionally reported that foundry workers exposed to PAH, who are at increased risk of developing lung cancer, generally showed higher levels of peripheral lymphocyte PAH–DNA adduction than controls; three out of the 12 exposed workers with increased peripheral PAH–DNA adducts also showed raised serum levels of *ras* or *fes* oncoproteins (but smoking may have been a confounding factor here). Although no quantitative relationship between BPDE–DNA adducts and parameters of proto-oncogene activation has yet been established, it is acknowledged that this strategy of biomarker analysis is a promising approach to the evaluation of lung cancer risk as a consequence of environmental exposure to carcinogens.

Conclusions

Estimating the risk of cancer associated with environmental exposure, such as that for respiratory tract cancer resulting from inhalatory intake of ubiquitous carcinogenic PAHs, involves many assumptions and uncertainties. Molecular cancer epidemiology is considered to be an innovative method by which to describe the relationship between exposure of humans to environmental carcinogens and the development of malignant neoplasias, particularly because of the advantages this approach may offer, as regards uncertainty reduction, over the classical epidemiological methods and studies in laboratory animals currently used to assess human cancer risk. The early euphoria that greeted the first generation of cytogenetic biomarkers was rather misplaced, because they do not seem to be directly associated with either cancerous events or endogenous carcinogen doses at target sites, as shown by PAH–DNA adduct analysis. Determination of DNA adduct levels in peripheral lymphocytes may be a sensitive method by which to assess intracellular carcinogen doses, but the relation between the kinetics of inhalatory intake of carcinogens from ambient air and the ultimate formation of peripheral lymphocyte DNA adducts has to be regarded with care. Measurements of markers of proto-oncogene activation in humans exposed to environmental carcinogens are now feasible. These second generation biomarkers offer two advantages—the analytical advantages of molecular biology techniques and an obvious causal link to carcinogenic processes.

Acknowledgement

I wish to thank the Ciba Foundation for the kind invitation to this meeting and for providing me with the opportunity to present my thoughts on this topic.

References

Ashby J 1988 Comparison of techniques for monitoring human exposure to genotoxic chemicals. Mutat Res 204:543–551
Balmain A, Brown K 1988 Oncogene activation in chemical carcinogenesis. Adv Cancer Res 51:147–182
Bond JA, Wallace LA, Osterman-Golkar S, Lucier GW, Buckpitt A, Henderson RF 1992 Assessment of exposure to pulmonary toxicants: use of biological markers. Fundam Appl Toxicol 18:161–174
Collins JF, Brown JP, Dawson SV, Marty MA 1991 Risk assessment for benzo(a)pyrene. Regul Toxicol Pharmacol 13:170–184
Dittrich KA, Krugh TR 1991 Mapping of BPDE adducts to human c-Ha-*ras1* protooncogene. Chem Res Toxicol 4:277–281
Goldring JM, Lucier GW 1990 Protein and DNA adducts. In: Hulka BS, Wilcosky TC, Griffith JD (eds) Biological markers in epidemiology. Oxford University Press, New York, p 78–104
Hulka BS, Wilcosky TC, Griffith JD (eds) 1990 Biological markers in epidemiology. Oxford University Press, New York
Kleinjans JCS, Albering HJ, Marx A et al 1991 Nitrate contamination of drinking water. No indication for genotoxic risk in human populations. Environ Health Perspect 94:189–193
Marx J 1991 Zeroing in on individual cancer risk. Science 253:612–616
Paleologo M, Van Schooten FJ, Pavanello S et al 1992 Detection of BPDE–DNA adducts in white blood cells of psoriatic patients treated with coal tar. Mutat Res 281:11–16
Perera F, Mayer J, Jaretzki A et al 1989 Comparison of DNA adducts and sister chromatid exchange in lung cancer cases and controls. Cancer Res 49:4446–4451

Perera F, Mayer J, Santella RM et al 1991 Biologic markers in risk assessment for environmental carcinogens. Environ Health Perspect 90:247–254

Preussmann RAW 1985 The role of food contaminants and additives in human carcinogenesis. In: Joossens JV, Hill MJ, Gebroers J (eds) Diet and human carcinogenesis. Excerpta Medica, Amsterdam, p 35–48

Rosenkranz HS, Ennever FK, Klopman G 1990 Relationship between carcinogenicity in rodents and the induction of sister chromatid exchanges and chromosomal aberrations in Chinese hamster ovary cells. Mutagenesis 5:559–571

Rothman N, Poirier MC, Baser ME et al 1990 Formation of polycyclic aromatic hydrocarbon–DNA adducts in peripheral white blood cells during consumption of charcoal-broiled beef. Carcinogenesis 11:1241–1243

Slebos RJC, Hruban RH, Dalesio O, Mooi WJ, Offerhaus GJA, Rodenhuis S 1991 Relationship between K-*ras* oncogene activation and smoking in adenocarcinoma of the human lung. J Natl Cancer Inst 83:1024–1027

Slooff W, Janus JA, Matthijssen AJCM, Montziaan GK, Ros JPM 1989 Integrated criteria document PAH. Report No. 758474011. National Institute of Public Health and Environmental Protection, Bilthoven

Sorsa M, Ojajärvi A, Salomaa S 1990 Cytogenetic surveillance of workers exposed to genotoxic chemicals. Teratog Carcinog Mutagen 10:215–221

Van Schooten FJ, Van Leeuwen FE, Hillebrand MJX et al 1990 Determination of benzo(a)pyrene diol epoxide DNA adducts in white blood cell DNA from coke oven workers: the impact of smoking. J Natl Cancer Inst 82:927–933

Van Schooten FJ, Hillebrand MJX, Van Leeuwen FE, Van Zandwijk N, Jansen HM, Kriek E 1992 Polycyclic aromatic hydrocarbon-DNA adducts in white blood cells from lung cancer patients: no correlation with adduct levels in lung. Carcinogenesis 13:987–993

WHO 1987 Air quality guidelines for Europe. WHO Regional Publications, Copenhagen (Eur Ser 23)

DISCUSSION

de Haan: You dealt with polycyclic aromatic hydrocarbons, mainly. Does cadmium act as a carcinogen?

Kleinjans: There are some suggestions that cadmium is carcinogenic to the kidney and the prostate gland, but it is not an acknowledged carcinogen in humans.

Klein: It's important to distinguish between experimental findings in the sort of studies you have been doing and the development of disease. The experiment you mentioned is based on the analysis of white blood cells, but the cancer you wish to look at may be of a quite different tissue—lung or liver, for example. What is the relationship between white blood cell DNA–PAH adducts or chromosome aberrations and the incidence of oncogene expression in cells of the organ of interest?

Kleinjans: This is of course the major problem at this time. We cannot find increased levels of PAH–DNA adducts in white blood cells from smokers, yet we find increased rates of chromatid exchanges, for example, in their white blood cells. We have to conclude that the carcinogen reaches the white blood cell DNA, but that something else then happens which we don't understand. White blood cell DNA is really the only sort of DNA that a scientist can get from an ambulant person. It is a surrogate source, and we have to be cautious.

Klein: At my research institute similar studies are being done on the relationship between radiation and chromosomal aberrations in white blood cells. We have a programme training doctors in Russia to apply these methods to help them to establish dose–response relationships in people from the area around Chernobyl. The problem is that white blood cells do not have a 'memory' of more than a couple of months.

Elliott: Can I ask about the reliability and reproducibility of these markers? Is it an all-or–none phenomenon? Will a foundry worker with *ras* proteins in his serum still have them a year later, or is there a distribution of effects?

Kleinjans: This is obviously a relevant question, but it's too early to give you an answer. If those proteins come from a spot inside the lung which is developing into a cancer, if there's a leakage of these proteins from these cells, it is unlikely that the proteins would reach the serum in amounts sufficient for them to be detected in the circulation. But those are the data. However, these are the first available results, so they have to be thoroughly studied before we go out into the field and measure oncogene expression in people.

Elliott: Have control sera been looked at? Are there people with these adducts who don't smoke or work in foundries?

Kleinjans: Analysis of oncogene expression in matched controls has not been done in this particular case, but it has been done in patients with lung cancer. Proteins encoded by oncogenes can be found in serum from patients who have lung cancers but do not smoke. The serum concentration of such proteins is really a marker for the carcinogenic process, not for the smoking.

James: Children's respiratory systems have high reactivity. I assume that you are using these DNA markers in an attempt to look at the long-term carcinogenic process. A completely different approach is taken at the TNO Toxicology Institute in The Netherlands, where they are looking at the response of bronchial mucosa, both in culture and *in vivo*, to air constituents, and are beginning to show differential bronchospasm in response to the synergistic effects of two types of aerosol. Won't this actually be more important than a possible effect on the DNA, particularly as in Europe there is an age-specific increase in, for example, asthmatic problems in children, for which there seems to be no reasonable explanation.

Kleinjans: With this specific problem in children it's probably more relevant to study lung functioning in relation to air pollution, which is actually being done by researchers in the Department of Epidemiology in my university, who are going out to schools during periods when air pollution is high and measuring children's lung functioning. The DNA techniques I spoke of have been developed to measure long-term effects, not acute, short-term effects. They cannot be used, for example, to look at the health effects of smog. We are dealing with long-term risks, and specifically the onset of cancer.

Hautvast: The environment will sometimes attack everybody. For example, if arsenic were abundant in people's immediate environments, everyone would probably die. However, many people smoke, but not all of them develop lung cancer. Which are the subgroups in our society who can cope with a particular environmental change? Is it the poor, the rich, the fat or the thin or whatever?

James: I believe a particular isoform of glutathione-*S*-transferase is more common than usual in those smokers who succumb to cancer (van Poppel et al 1991). There is new evidence showing that this particular isoform is important in modulating carcinogen production. Thus, the individual sensitivity of people is actually being entrained by their specific inheritance of these metabolic pathways.

Kleinjans: That's correct, and it is also true for the cytochromes which oxidize the chemicals, and probably also for the activity of the DNA repair system, which is a multienzymatic system which can be modulated by dietary factors. The operating status of the DNA repair system can differ from person to person, being modulated partly by genetic aspects, and partly by dietary or health aspects.

James: Has it really been shown that DNA repair mechanisms are modulated by diet?

Kleinjans: There has been one study, and we are doing the second one.

Elliott: It is for cardiovascular diseases that we have the most considerable knowledge about the major risk factors. We know that whole populations are at risk of premature morbidity and mortality from cardiovascular diseases because of high prevalance of smoking and high average blood pressures and serum cholesterol. Thus, for a non-smoking Westerner with blood pressure lower than average for that population and lower than average cholesterol, the most likely cause of death is coronary heart disease. This illustrates the concept of population risk versus individual risk. Although we are good at identifying *groups* at increased risk of coronary heart disease or lung cancer, with *individuals* we are generally not so good, with one or two exceptions (those with familial hypercholesterolaemia, for example). The approach to prevention needs to involve a population-wide strategy which recognizes the concept of the sick population, i.e., as a group, we eat too much, we smoke too much, and our blood pressures are too high.

Hautvast: If, for example, I eat an apple and develop an allergy in my mouth, I will know, but if I eat three eggs in one day I don't know whether my blood cholesterol increases. In some people, the hyper-responders, it will increase, whereas in others there will be hardly any change. We often consider all human beings as being biologically equal, in a community approach. This is a justified approach for health programmes.

James: The implication of that is that you ought to find an appropriate genetic marker, then use those individuals carrying the marker gene to test the sensitivity of susceptible individuals to the agent.

Bobrow: One of the concepts in radiation protection, for example, is to look at critical groups—people who are most highly exposed. They are a sort of worst case scenario. This is a reasonable way of trying to set standards which will protect the large bulk of the population. What you suggested works in the opposite way. For example, if you screen large populations for genetic markers to identify subgroups who can get away with smoking (which is likely to be technically feasible in the next year or two), you will find yourself in a situation where the knowledge may be counter-productive. Much of the discussion so far has concerned situations where we actually know quite a lot about how to prevent ill health but where the problem lies in the application of that knowledge: really quite basic knowledge is not being applied. Do we really want to add to the complexity of legislative and social machinery by setting different standards for certain subgroups of the population?

Hautvast: This will be the future, whether we like it or not. It will be possible to decide whether a particular person ought to eat something.

Reference

van Poppel G, de Vogel N, van Bladeren PJ, Kok FJ 1991 Increased cytogenetic damage in smokers deficient in glutathione-S-transferase isozyme μ. In: Annual report of the TNO-CIVO Toxicology and Nutrition Institute. Zeist, The Netherlands, p 47

Radiation-induced disease

Martin Bobrow

Division of Medical and Molecular Genetics UMDS, Paediatric Research Unit, 8th Floor Guy's Tower, Guy's Hospital, London Bridge, London SE1 9RT, UK

Abstract. The term radiation covers a wide spectrum of forms of energy, most of which have at one stage or another been suspected of causing human ill health. In general, study of the effects of radiation on health involves a mix of scientific disciplines, from population epidemiology to physics, which are seldom if ever found in a single scientist. As a result, interdisciplinary communication is of the utmost importance, and is a potent source of misunderstanding and misinformation. The forms of radiation which have been most specifically associated with health effects include ionizing and ultraviolet radiation. Claimed effects of electromagnetic and microwave radiation (excluding thermal effects) are too indefinite for detailed consideration. Ionizing radiation is a well-documented mutagen, which clearly causes cancers in humans, and human exposure has been increased by atomic weapons testing and medical and industrial uses of radioactivity. There is also a growing awareness of the possible role of some types of natural radiation, such as radon, in causing disease. Ultraviolet radiation is also associated with cancers, and is suspected of involvement in the increasing incidence of skin cancers in European populations. Factors thought to underlie recent changes in exposure to these mutagens are discussed.

1993 Environmental change and human health. Wiley, Chichester (Ciba Foundation Symposium 175) p 182–196

Radiation-induced disease is a highly emotive subject, and political and public perception of the field is such that it is difficult to consider the objective evidence independently of this. There are few fields of science so influenced by newspapers, and it is arguable that study of the communication gap between scientists and the rest of society in this area is as important as the health effects themselves. If we could understand the evolution of the public debate on this subject, we might be better placed to understand the substantial public mistrust of scientific methodology and of scientists themselves. Here, I shall briefly review the current evidence on human disease caused by radiation.

'Radiation' means spreading from a source, and although the term is commonly thought of as referring to emissions from radioactive substances, it also encompasses several other types of energy. There are particulate radiations, such as α and β particles and neutrons, and wave-forms of energy, such as light. Spectral radiation covers a wide range of frequencies and wavelengths, from very long-wave radio, through microwaves and the visible light spectrum to ultraviolet light, and then the so-called ionizing radiations, which at the shorter wavelengths include γ and cosmic rays. These radiations, depending on their physical characteristics and energy levels, can penetrate living tissue to varying extents, depositing their energy as they do so. This is the basis for their harmful effects. Virtually all forms of radiation have at one time or another been accused of having ill effects on people.

Non-ionizing radiation

Electromagnetic energy

Electromagnetic radiation (e.g., from power lines) has been associated with cancers, particularly brain tumours and leukaemias, but the evidence is conflicting and many of the studies are flawed (National Radiological Protection Board [NRPB] 1992). Experiments investigating cellular effects of electro-magnetic energy have failed to reveal biologically plausible pathways to lend weight to the epidemiological investigations. Scientists are always at their worst in coming to firm negative conclusions, but there is currently little persuasive evidence that exposure to this energy source is harmful. This absence of evidence has not inhibited legal action in the USA, nor has it deterred expensive avoidance of unproven hazards (Florig 1992). With increasing dependence on electric power, devices such as VDUs (video display units) which generate electro-magnetic fields, and the use of medical imaging techniques such as MRI (magnetic resonance imaging), public exposure to electromagnetic radiation seems likely to increase. Whether or not this is important depends on whether or not there is any real hazard from such exposure, and further research is therefore needed.

'Optical' wavelengths

This range of wavelengths contains an important and well-proven human carcinogen, ultraviolet light. UV is conventionally divided into three ranges, UVA (315–400 nm), UVB (280–315 nm) and UVC (100–280 nm) (International Commission on Illumination & International Electrotechnical Commission 1987). The shorter wavelengths have little penetrating power (indeed, the shortest are not propagated through air), but UVA and UVB are present in sunlight, and can induce mutations in DNA, particularly formation of thymidine dimers. Work on mice suggests that the spectrum of UV causing non-melanoma skin cancers is similar to that causing acute erythema (Cole et al 1986).

There is considerable epidemiological evidence linking UV radiation with skin cancer (UVB, in particular, is associated with both squamous and basal cell skin cancers). The reported incidence of these cancers has recently increased sharply in the UK and Europe. For example, the total number of registered cases of non-melanotic skin cancers in England and Wales increased from 18 891 in 1976 to 26 767 in 1986 (Office of Population Censuses and Surveys [OPCS] 1981, 1991a), an increase of 42%. This has been attributed to increasing exposure to sunlight resulting from changing social, holiday and sun-tanning habits which have coincided with improved transport, making annual holidays in sunny areas available to large numbers of north Europeans.

Skin cancer is the second most common cancer in the UK, and although treated non-melanoma skin cancers have a good prognosis, the morbidity and disfigurement caused constitute a considerable public health problem. Malignant melanoma is rarer, but has a high associated mortality. The continuing rise in skin cancers (see Fig. 1) has been targeted as a major problem in *The Health of the Nation* (Department of Health 1992), which sets out the Government's strategy for health in England. The total number of registered cases of malignant melanoma in England and Wales increased from 1675 in 1976 to 2802 in 1986 (OPCS 1991b), an increase of 67%, and in Scotland they increased from 257 in 1979 to 469 in 1989, an increase of 82% (MacKie et al 1992).

Non-melanoma skin cancers appear to be related to total cumulative UV exposure. For malignant melanoma, the relationship to UV is widely accepted but less clear cut; neither the geographic nor the anatomical distribution (Pathak 1991) of lesions suggests the simple relationship which appears to hold for non-melanoma skin cancers (Slaper & van der Leun 1987). The currently favoured hypothesis is that tumour initiation is related to intermittent acute exposure in the early years of life (Armstrong 1988).

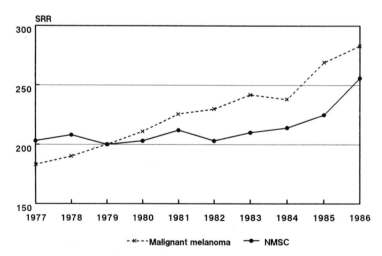

FIG. 1. Skin cancer registrations in England and Wales between 1977 and 1986. Standardized registration ratios (SRR) for persons aged under 75 (base year 1979) are plotted. NMSC, non-melanoma skin cancer. Data derived from OPCS 1991b.

Skin cancer is therefore increasingly common, and the rise is highly likely to be related to exposure to UV radiation. The incidence will continue to rise unless effective action is taken. The social habit of sun-tanning may not decrease in popularity. There has also been much discussion of a postulated reduction in stratospheric ozone. Ozone acts as a filter of solar UV and if it is indeed depleted one long-term effect could be an increase in ground-level exposure to UV radiation. This change is likely to be a slow process, however: the northern hemisphere at UK latitudes has lost only about 5% of its stratospheric ozone over the last decade. The increased amount of UV radiation to which the average UK citizen is exposed as a result of this change is still considerably less than that encountered during a fortnight's Mediterranean beach holiday. However, the process is potentially cumulative, difficult to reverse and will affect very large numbers of people. The modest magnitude of the changes observed to date should not be used as an excuse for inaction, although it is a good argument for considered and effective action rather than short-term panic measures.

Ionizing radiation

Natural ionizing radiation

Most people's greatest exposure to ionizing radiation comes from natural rather than man-made sources, from radioactivity in food-stuffs and the earth itself, and from cosmic rays (Fig. 2). A significant component of background radiation derives from the natural gas radon. Its potential effects on health have been

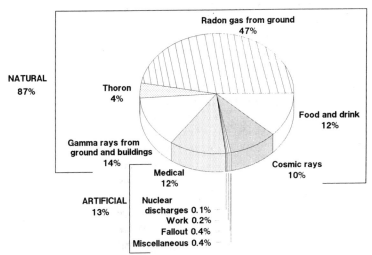

FIG. 2.   Relative contribution of various natural and artificial sources of radiation to the average annual dose of radiation to the UK population. Data derived from National Radiological Protection Board (NRPB) 1989a.

widely discussed over the past few years, because of their possible magnitude and also because of their geographical variability and the possibility of reducing exposure of those most at risk at relatively modest cost.

Radon is a decay product of natural uranium, found in many rocks, particularly granites. In very fractured granites the gas finds its way to the surface relatively rapidly, and thence into dwellings. Virtually all the non-occupational exposure of humans is believed to occur within houses. Heated dwellings at slightly lower than external pressure draw gas in; its access to the interior depends on floor construction and poor ventilation encourages its retention. Exposure is therefore determined not only by local geology, but also by local building practices.

Radon is a short-lived α-emitter, and, being a gas, exposure is mainly by inhalation. Because radon is fat-soluble, it has been argued that it may become localized in bone marrow, causing leukaemia (Henshaw et al 1990), and epidemiological observations have been interpreted as suggesting a correlation between leukaemia rates and background radon levels (Henshaw et al 1990); however, these conclusions are controversial.

Uranium miners subjected to heavy occupational exposure to radon clearly show a raised incidence of lung cancer, and the effect is synergistic with that of tobacco smoke (Committee on the Biological Effects of Ionizing Radiation [CBEIR] 1988). Only very recently have results from sizeable studies appeared demonstrating a relationship between radon exposure and lung cancer in low dose-rate non-occupational settings (Pershagen et al 1992). In the UK the heaviest exposure to radon is in parts of Cornwall, but the county has low lung cancer rates. However, this negative finding is inconclusive, because lung cancer rates are dominated by smoking to such an extent that other effects may be masked, and Cornwall's unusual demographical characteristics complicate controlled studies. Some workers doubt that radon has any important public health effects (Bowie & Bowie 1991), but this is a minority view. Large, carefully controlled studies currently under way should resolve this issue.

Man-made ionizing radiation

The final category of radiation considered here is man-made ionizing radiation. High doses of ionizing radiation cause cell death. The early acute effects of exposure are massive damage to the gastrointestinal mucosa, skin and bone marrow, as seen in those exposed to atomic bombings, those clearing the Chernobyl explosion site, and in other victims of nuclear accidents.

All forms of ionizing radiation are known to be mutagenic. In principle, these mutations will give rise to cancers (somatic cell mutation) or to hereditary effects (germ cell mutation). Radiation's carcinogenic, and in particular its leukaemogenic, potential has been demonstrated in many well-validated studies on, for example, people given radiotherapy for ankylosing spondylitis or other disorders (Spiess et al 1989), children exposed to X-rays *in utero* (Gilman et al 1988), the

Japanese survivors of the atomic bombs (United Nations Scientific Committee on the Effects of Atomic Radiation [UNSCEAR] 1988), or painters of luminous dials (Finkel et al 1969). These data cover both acute and chronic exposure, and there is also a vast amount of confirmatory animal data (CBEIR 1990).

Hereditary effects have been shown in animals, but not directly in humans, probably because of the difficulty in collecting sufficient data for their formal demonstration. There seems no reason to doubt that human exposure to ionizing radiation will lead to both cancer and inherited disease, but how much radiation, of what form and at what rate of exposure will produce a particular amount of detriment is presently unknown.

Because of the very large potentially exposed populations, the most important public health issue is that of chronic exposure to relatively low doses at low dose rates. Various types of chronic low dose and low dose rate exposure to ionizing radiation have been studied, for example, fall-out from weapons testing, exposure of local populations in the vicinity of nuclear installations, and nuclear accidents, of which the most dramatic was that at Chernobyl. No immediate health effects are observable from these relatively low doses, making effective monitoring difficult.

Studies of the effects of chronic low-dose radiation involve assessment of both population dosimetry and epidemiological evidence of health effects, but both have inherent difficulties. It is not possible to measure doses received by large numbers of people over long periods of time accurately. External radiation exposure is easier to assess than internal irradiation, but the latter is likely to be more important. The particular chemical form of internal irradiation may also be crucial, because different chemicals will be localized in different tissues through physiological mechanisms. One of the important potential forms of internal exposure to irradiation is that from α-emitters, such as plutonium, which deposit energy in a very dense fashion within a few μm in body tissues. The crucial event in carcinogenesis is whether an ionization track damages a sensitive stem cell, so that the difference between a substance which naturally becomes concentrated close to stem cells and one which is concentrated a few millimetres away may be critical. However, we do not know enough about where stem cells are located, nor do we know enough about uptake rates, tissue localization, or rates of turnover of important radionuclides.

Epidemiological studies also have methodological difficulties. Defining the precise areas, time periods and diseases to be studied produces multiple hypotheses, confounding the assessment of statistical significance. With emission from a local source such as a nuclear installation, local residents will probably be the most highly exposed, but the numbers will be extremely small for an epidemiological study. Larger populations are more satisfactory statistically, but may dilute real exposure effects. Direct observation of effects on health requires accurate monitoring of large populations over long periods, which is extremely difficult. The UK studies have been possible only because of a long-standing national cancer registration system, which is particularly well developed with regard to childhood cancers (Draper 1991).

Empirical data do not clearly show changes in cancer rates related to fall-out levels (Darby et al 1992, Modan 1992), apart from in small populations exposed to high levels of radiation because of proximity to test sites—for example, soldiers who were present at nuclear tests (Darby et al 1988). The epidemiological effects of the Chernobyl accident on wider populations have not yet been unequivocally demonstrated. The very large exposed populations of Europe would have received low doses, the effects of which would be difficult to disentangle from background rates, whereas the more intensely exposed local populations were not previously monitored in a way which set a reliable base-line against which to confidently identify any increases. The question is not whether such harm has occurred; the problem is that it is difficult to quantify, frustrating the formulation of public policy.

Much recent interest has focused on studies of populations living in the vicinity of nuclear installations. In 1983 a television producer, James Cutler, drew attention to an excess of childhood leukaemia in the vicinity of the large Sellafield reprocessing works in Cumbria in the UK; the validity of this observation has since been throroughly confirmed, although its interpretation remains problematical (see Table 1; see also Gardner et al 1990). Similar studies initiated near Britain's only other reprocessing plant, Dounreay, revealed an excess of the same disorder (Heasman et al 1986, Urquhart et al 1991)—too much of a coincidence to be dismissed, notwithstanding the small numbers of cases involved in these relatively underpopulated regions. Studies of cancer incidence around other UK nuclear sites have yielded some positive (Roman et al 1987) and some negative results (Ewings et al 1989), but none of these was a reprocessing plant and the potential exposure, both public and occupational, would be quite different. There have also been a number of negative findings from studies outside the UK (e.g., Hill & Laplanche 1990), to which similar reservations about the comparability of data apply.

TABLE 1   Mortality to 30.6.86 among children born 1950–1983 to mothers resident in Seascale civil parish[a]

| Cause of death | No. of deaths | | | 95% Confidence interval |
	O	E	O/E	
Leukaemia	5	0.53	9.36	3.04–21.84
Non-Hodgkin's lymphoma	1	0.12	8.45	0.21–47.1
All malignancies	9	1.6	5.63	2.58–10.69
Total deaths	27	32.3	0.84	0.55–1.22

[a]Seascale is the closest village to the Sellafield Reprocessing Plant.
Modified from Gardner et al 1987.
O, observed; E, expected.

The excess of leukaemia in young people around Sellafield and Dounreay has been confirmed by several studies, but not explained. Several hypotheses have been advanced.

(1) The disorder is the direct result of exposure of local populations to emissions or discharges from the plant. Confirmation of this obvious starting hypothesis is confounded by the difficulty of reconciling estimates of local dosimetry satisfactorily with observed health effects and what is known of radiation dose–response relationships. These dosimetric arguments soon achieve appalling complexity, and although the weight of evidence is against a direct effect there are still contrary opinions.

(2) Gardner et al (1990) reported a relation between leukaemia and paternal external radiation dose, and hypothesized that childhood leukaemia may follow paternal mutation. This has not been confirmed (or refuted) and must be regarded as controversial. Several other possible connections can be postulated between childhood leukaemia and parental occupation, and these are well worth further investigation.

(3) Elevated leukaemia rates could be related to demographic features, rather than to the nuclear installation itself. It has, for example, been observed that leukaemia rates are higher in areas considered as being suitable for the construction of nuclear power stations but where they were not in fact built (Cook-Mozaffari et al 1989). Kinlen has amassed evidence that childhood leukaemia may be related to 'herd immunity' effects of population migrations, which may particularly affect remote regions in which industrial complexes are built. This effect would be causally related to a postulated infective/immunity basis for leukaemia, rather than radiation (Kinlen 1988, Kinlen & Hudson 1991).

The problem with a review of this field is that it is easy to be overly sceptical—enough is known about the physical properties and biological effects of ionizing radiation to enable virtually any study to be decried, so that the main issues are easily obscured in a welter of fact and counter-fact. It is certain that all forms of ionizing radiation are mutagenic and carcinogenic, and there is no evidence for a threshold below which these effects disappear, so it can be assumed that all exposure is harmful. What is disputable is the *quantitation* of this phenomenon, particularly in relation to the very low dose and dose rate exposures likely to affect large populations. Much less is known about the effects of internal exposure to particulate radiation than the effects of penetrating forms such as X-rays and γ-rays.

It is possible for an optimist to believe that studies of the elevated incidence of cancer in the vicinity of some nuclear installations may eventually lead to new understanding of the mechanisms of causation of leukaemia in young people. A better understanding of the process of leukaemogenesis in general is likely to assist in interpretation of the true risks associated with very low dose radiation. Understanding the biology of the disease will bring benefits well beyond the narrow field of radiation risk estimation. These are all powerful arguments for further detailed research into these fascinating and still unexplained phenomena. Although the number of deaths identified in the areas studied is small, they are potentially avoidable, and if their origin is not understood, their numbers are likely to increase rather than diminish.

It is impossible to consider the effects of radiation on human health without being aware of the unusually emotive public image of this subject. Why does radiation engender such an unusually acute public reaction? There are a few obvious reasons. Firstly, nuclear energy was initially developed and used for the construction of weapons of mass destruction, and the entire nuclear industry has relatively recently evolved from this awesome military background. The public memory will long associate the word 'radiation' with mass death. Secondly, our senses do not warn us of the presence of ionizing radiation. This is particularly worrying because its effects can spread widely in space, crossing national boundaries, and personal defence against them is very difficult. The fact that major effects are delayed in time is also a worry, because it is thus difficult to know that the danger has passed and that one has 'escaped'. Finally, there is a general (and reasonable) intolerance of hazard which is imposed by others without immediate choice by or perceived benefit to the individual, as compared with far more hazardous activities such as driving and smoking which are voluntarily undertaken.

There are large vested interests involved on all sides of these arguments. Scientists with good credentials, arguing opposite points of view on highly technical matters in public, are hardly likely to increase the prospect of the general public concluding that the 'experts' really know what they are talking about.

Should nuclear power ever again be used in war, the devastation would clearly be on a scale beyond the issues under discussion today. Even without this, increasing demands for energy and increasing technical ingenuity will mean that human populations will continue to be exposed to radioactivity, although the magnitude of this may diminish further with improved containment procedures. The average dose equivalent in the UK from both external irradiation and intake of radionuclides is about 5 μSv a year from fall-out from nuclear weapons tests at present, in comparison with the maximum dose, about 140 μSv, that was received in 1963 (NRPB 1989b). Whether peaceful uses of nuclear energy, particularly for power generation, can or should be radically reduced is an enormously important and difficult political issue. If it can be resolved by informed debate and rational analysis, rather than by short-term reactions to sudden media and political pressures, then society will have come a long way towards scientific maturity.

References

Armstrong BK 1988 Epidemiology of malignant melanoma: intermittent or total accumulated exposure to the sun? J Dermatol Surg Oncol 14:835-849

Bowie C, Bowie SHU 1991 Radon and health. Lancet 337:409–413

Cartwright RA, Alexander FE, McKinney PA, Ricketts TJ 1990 Leukaemia and lymphoma: an atlas of distribution within areas of England and Wales, 1984–88. Leukaemia Research Fund, London

Cole CA, Forbes PD, Davies RE 1986 An action spectrum for photocarcinogenesis. Photochem Photobiol 43:275–284

CBEIR 1988 BEIR IV Report: health risks of radon and other internally deposited alpha-emitters. National Academy Press, Washington, DC

CBEIR 1990 BEIR V Report: health effects of exposure to low levels of ionizing radiation. National Academy Press, Washington, DC

Cook-Mozaffari P, Darby S, Doll R 1989 Cancer near potential sites of nuclear installations. Lancet 2:1145–1147

Darby SC, Kendall GM, Fell TP et al 1988 Mortality and cancer incidence in UK participants in atmospheric nuclear weapons tests and experimental programmes. HMSO, London (NRPB-R214)

Darby SC, Olsen JH, Doll R et al 1992 Trends in childhood leukaemia in the Nordic countries in relation to fallout from atmospheric nuclear weapons testing. Br Med J 304:1005–1009

Department of Health 1992 The health of the nation. A strategy for health in England. HMSO, London

Draper G (ed) 1991 The geographical epidemiology of childhood leukaemia and non-Hodgkin lymphomas in Great Britain, 1966–83. HMSO, London (Stud Med Popul Subj 53)

Ewings PD, Bowie C, Phillips MJ, Johnson SAN 1989 Incidence of leukaemia in young people in the vicinity of Hinkley Point nuclear power station, 1959–86. Br Med J 299:289–293

Finkel AJ, Miller CG, Hasterlik RJ 1969 Radium-induced malignant tumours in man. In: Mays CW, Gee WSS, Lloyd RD, Stover BJ, Dougherty JH, Taylor GN (eds) Delayed effects of bone-seeking radionuclides. University of Utah Press, Salt Lake City, UT, p 195–225

Florig HK 1992 Containing the costs of the EMF problem. Science 257:468–492

Gardner MJ, Hall AJ, Downes S, Terrell JD 1987 Follow up study of children born to mothers resident in Seascale, West Cumbria (birth cohort). Br Med J 295:822–827

Gardner MJ, Snee MP, Hall AJ, Powell CA, Downes S, Terrell JD 1990 Results of case-control study of leukaemia and lymphoma among young people near Sellafield nuclear plant in West Cumbria. Br Med J 300:423–429

Gilman EA, Kneale GW, Knox EG, Stewart AM 1988 Pregnancy X-rays and childhood cancers: effects of exposure age and radiation dose. J Radiol Prot 8:3–8

Heasman MA, Kemp IW, Urquhart JD, Black R 1986 Childhood leukaemia in northern Scotland. Lancet 1:266

Henshaw DL, Eatough JP, Richardson RB 1990 Radon as a causative factor in induction of myeloid leukaemia and other cancers. Lancet 1:1008–1012

Hill C, Laplanche A 1990 Overall mortality and cancer mortality around French nuclear sites. Nature 347:755–757

International Commission on Illumination (CIE), International Electrotechnical Commission (IEC) 1987 International lighting vocabulary. IEC, Geneva

Kinlen L 1988 Evidence for an infective cause of childhood leukaemia: comparison of a Scottish new town with nuclear reprocessing sites in Britain. Lancet 2:1323–1327

Kinlen LJ, Hudson C 1991 Childhood leukaemia and poliomyelitis in relation to military encampments in England and Wales in the period of national military service, 1950–1963. Br Med J 303:1357–1362

MacKie R, Hunter JAA, Aitchison TC et al 1992 Cutaneous malignant melanoma, Scotland, 1979–89. Lancet 339:971–975

Modan B 1992 Low dose radiation carcinogenesis. Eur J Cancer 284:1012–1013

NRPB 1989a At a glance series. Radiation doses—maps and magnitudes. NRPB, Chilton, Didcot

NRPB 1989b Living with radiation, 4th edn. HMSO, London

NRPB 1992 Electromagnetic fields and the risk of cancer. Report of the advisory group on non-ionizing radiation. NRPB, Chilton, Didcot (Doc NRPB vol 3, no 1)

OPCS 1981 Table 2, registration of newly diagnosed cases of cancer: sex, site and age, 1976. In: Cancer statistics registrations. MBI 7. HMSO, London

OPCS 1991a Table 2, registration of newly diagnosed cases of cancer: sex, site and age, 1986. In: Cancer statistics registrations. MBI 19. HMSO, London

OPCS 1991b Table 1, Standardized registration ratios for persons aged under 75 (base year 1979): sex and site, 1977–1986. In: Cancer statistics registrations. MBI 19. HMSO, London

Pathak MA 1991 Ultraviolet radiation and the development of non-melanoma skin cancer: clinical and experimental evidence. Skin Pharmacol 4 (suppl 1):85–94

Pershagen G, Liang ZH, Hrubec Z, Svensson C, Boice JD Jr 1992 Residential radon exposure and lung cancer in Swedish women. Health Phys 63:179–186

Roman E, Beral V, Carpenter L et al 1987 Childhood leukaemia in the West Berkshire and Basingstoke and North Hampshire District Health Authorities in relation to nuclear establishments in the vicinity. Br Med J 294:597–602

Slaper H, van der Leun JC 1987 Human exposure to ultraviolet radiation: quantitative modelling of skin cancer incidence. In: Passchier WF, Bosnjakovic BFM (eds) Human exposure to ultraviolet radiation: risks and regulations. Elsevier Science Publishers, Amsterdam, p 155–171

Spiess H, Mays CW, Chmelevsky D 1989 Malignancies in patients injected with radium 224. In: Taylor DM, Mays CW, Gerber GB, Thomas RG (eds) Risks from radium and thorotrast. British Institute of Radiology, London (Br Inst Radiol Rep 21) p 7–12, 27

UNSCEAR 1988 Annex F: sources, effects and risks of ionizing radiation. In: Radiation carcinogenesis in man. United Nations, New York

Urquhart JD, Black RJ, Muirhead MJ et al 1991 Case-control study of leukaemia and non-Hodgkin's lymphoma in children in Caithness near the Dounreay nuclear installation. Br Med J 302:687–692

DISCUSSION

Lake: I was hoping that we might derive at least a slightly beneficial effect from radon in the environment, through the mutations it induces enabling us to adapt more rapidly to our changing environment.

Bobrow: Evolution has been going for quite a long time now—I think we would have to get up very early in the morning to invent things which haven't yet been invented by Nature.

Klein: For many decades, radiation was considered a positive, healthy thing and was used in medicine. There was a turn around in public opinion, though its benefits still exist. You mentioned that radiation protection is, in a way, a problem of legislation. If the concepts of radiation protection had been set up 20–30 years ago, the results would be more irrational, because at that time people were willing to accept a certain risk from the natural environment. The concept of risk assessment in radiation protection is based on the additional risk in comparison with the natural background risk. Another part of the concept of radiation protection is the balance between the economic or technical measures which would be needed to avoid a certain level of damage and the resulting protective benefits. We would be much happier if drinking water quality regulations had been developed with such a rational approach to the problem of pollution.

You suggested that we have not learnt much from the Chernobyl accident. I am less pessimistic. This story is not yet closed. Current radiation protection guidelines are based on epidemiological studies in humans in Hiroshima and Nagasaki, and on experiments on animals, where there are problems in extrapolating from high dose to low dose and from animals to humans. These two approaches form the basis for the doses considered acceptable today. It's true that many years were wasted in the Soviet Union because of misinformation and restriction of access to data and the opportunity to couple the information on exposure of a large number of people to the pattern of health problems that developed was missed. Even now, six years after the event, we are still in a position in which we could combine our forces with those of the Russian doctors and epidemiologists to find out better ways of estimating the doses to which people were exposed and the likely effects on health. People in my institute and others in Germany and the USA are involved in programmes intended to improve dosimetry to help in decision-making, to distinguish between populations at risk and populations not at risk, and to find out the health consequences.

The occurrence of unexpected cases of cancer in young children or elderly people gives us the opportunity to find the gaps or the unknown factors in our risk assessment protocols. It's fair to say that some observations have been made which would not have been predicted on the basis of current concepts. We know there are weaknesses in our assessment systems, so this work will stimulate radiation protection research. We have to ensure that radiation protection research continues, so that there are institutes where there are people competent to deal with this important problem in the future.

Bobrow: You reminded us that radiation was thought to be beneficial; there are still people who believe that, and indeed it does have important uses. I have met several people who were the young radiation physicists and radiobiologists of the 1950s. They, at that time, must have been like the molecular biologists of today—they had come across one of *the* important scientific advances for the benefit of mankind, and they were filled with enthusiasm for this gift they had to give. It's been difficult for those people to have their gift thrown back in their faces in ways they regard as essentially irrational. They say that its military use is behind us, and that we should look at all the good it can do, then other people quibble about three cases of this, and four cases of that. We should not think these scientists were unable to see the consequences, and didn't think about the hazards. They were just as bright and had just as much insight as us—they lived then and we live now, and our successors will think about us in the same way that we think about them, unless we become a lot smarter than we have been until now and learn, not how to solve one problem in this complex interaction between man and the environment, but how to see three or four moves ahead to decide what will be the ill effects of our solution to today's problem.

Your second point was about the structure of radiation protection regulations. You are right that the legislative protection apparatus for radiation is extremely tight and precise in comparison with that for other environmental hazards. The point I was trying to make is not that the legislative apparatus is poorly set up. My point was about the way in which the UK Government, at least, takes political decisions about the sort of energy policy in which it should invest. Political and public reactions are not well correlated with numerical estimates of risk, as has been pointed out frequently (e.g., Royal Society Study Group 1983). The demonstrable local health effects of Sellafield's operations might number, if you are really pessimistic, some hundreds of deaths over 30 years, and there has been an expenditure of some hundreds of millions of pounds in order to lessen its environmental impact. I am strongly in favour of cleaning up discharges from dirty plants, but the serious cost–benefit analysis of these decisions is not always clear cut.

With Chernobyl, I think we have missed the opportunity to find out much about early effects such as childhood cancers, but perhaps you are right that we may learn something about later onset cancers.

Zehnder: You made some important points. Why should we actually bother about a few deaths when we have problems to deal with where hundreds of thousands are affected? But we *do* care. This is how society reacts, how we as people, react. To us, it's not the mass which counts, it's the individual. To see a child dying of leukaemia touches us much more than knowing that in another country 50 000 people will be wiped out. We have to be careful not to adopt administrator's attitudes, making decisions solely on the basis of numbers.

James: Rather than being an administrator, or a scientist, or simply a human being, you have to learn the art of being a controlled schizophrenic; you have to have the sensitivity of a public individual acting irrationally, and then you have to have the responsibility of a scientist. You are then put into an administrative position in which there are choices to be made in terms of public expenditure. The politician who makes a decision on the basis of his perception of the public's attitudes recognizes that he needs to be popular. Surely our role is to bring rationality to bear, yet at the same time understand the sensitivities of the issues such that we project the basis for that rationality in such a way that we can begin to change public perception.

Rabbinge: The perception of risk is of course the basis of all decision-making. It's not sufficient to say the decision-making and the policy set by administrators is irrational. In the early 1970s many researchers proposed that we should distinguish between macro risks and micro risks: a micro risk is a normal risk, when the event has a low probability of occurring and would have a small effect on society, whereas with a macro risk the likelihood is low but the effects would be great. An accident at a power plant is a typical example of a macro risk. Also, as pointed out earlier (p 190), you have to distinguish between the private risks an individual is prepared to accept and public risks. Private risks are things like drinking too much alcohol, smoking, or driving a car, and a public risk is something the individual is not voluntarily exposed to. How do you deal with these differences in perceptions, public risk and private risk, micro risks and macro risks?

Bobrow: I highlighted risk perception because I think it's a central issue, but of course I don't know the answers. With rare, and usually unpleasant, exceptions, politicians act as a sort of integrating mechanism, reacting to the pressures which are around them. Industry and the military, for example, are extremely adept at applying the kind of pressure that makes politicians do what they want them to do. Scientists are not generally as good at that, which is why we have a lot of trouble getting what we think is sensible into the political arena. I don't know the reason for this, but I can imagine a number of contributing factors. One is that we always ask questions—that's what science is about, whereas politics is about giving clear-cut answers. Whenever I write for a newspaper, I write something with lots of qualifying clauses, 'perhaps' and 'it is conceivable that in the event of', but when it's printed all the qualifications have gone, leaving clear black-and-white statements, because that's what the press, public and politicians want to read, and that's what gets results. In my worst moments, I think it is not even the journalists who really make public policy, but the sub-editors who write the headlines. They compress your extremely complex article into six words in big letters, which have more impact than anything else. Here we are, an intelligent group of people with clear-cut understanding (not terribly detailed in some respects but enough to be going on with) about nutritional problems, about infective problems, about all sorts of important health problems, but instead of getting out there and writing headlines and analysing how to make the politicians do what we want them to do, we argue in public. For every expert who says you should do one thing, there's another one who says you should do something else. They're usually arguing about small differences, but the effect is that the public think we're all mad, that we don't know what we're talking about, and they discount the whole lot of us.

Mansfield: You mentioned the physicists who worked on radiation in the 1950s. It's worth recording that they actually did a good publicity job. The public had confidence in their activities. The first nuclear power stations were opened in a blaze of glory, and even a big accident at Seascale didn't shake public confidence very much—it survived even that. Public confidence in scientists as a whole has now been shaken so much that molecular biologists are having great difficulty in persuading the public to accept simple genetically modified organisms, such as tomatoes that will not ripen too rapidly on the supermarket shelf—the public are terrified of those! Our powers of communication with the public must have deteriorated enormously over the past 40 years.

Reference

Royal Society Study Group (ed) 1983 Risk assessment: a study group report. Royal Society, London

Reproductive health, population growth, economic development and environmental change

Dennis W. Lincoln

MRC Reproductive Biology Unit, University of Edinburgh Centre for Reproductive Biology, 37 Chalmers Street, Edinburgh, EH3 9EW, UK

Abstract. World population will increase by 1000 million, or by 20%, within 10 years. Ninety-five per cent of this increase will occur in the South, in areas that are already economically, environmentally and politically fragile. Morbidity and mortality associated with reproduction will be greater in the current decade than in any period in human history. Annually, 40–60 million pregnancies will be terminated and 5–10 million children will die within one year of birth. AIDS-related infections, e.g. tuberculosis, will undermine health care in Africa (and elsewhere) and in places AIDS-related deaths will decimate the work-force. The growth in population and associated morbidity will inhibit global economic development and spawn new problems. The key issues are migration, the spread of disease, the supply of water and the degradation of land, and fiscal policies with respect to family planning, pharmaceuticals and Third-World debt. Full education, particularly of women, and more effective family planning in the South have the power to unlock the problem. Failure will see the developed countries, with their 800 million population, swamped by the health, economic and environmental problems of the South, with its projected population of 5400 million people for the year 2000.

1993 Environmental change and human health. Wiley, Chichester (Ciba Foundation Symposium 175) p 197–214

Opinions are highly polarized on the interrelationships between reproductive health, population growth, economic development and changes in the environment (Fig. 1). One school clings to a Malthusian view where growth in population inevitably outstrips the capacity of the earth to provide the necessary support, such that growth is finally brought into check by a reduction in fertility and an increase in disease, famine and conflict. Elements of this argument are evidenced by the fall in the human sperm count, the spread of human immunodeficiency virus (HIV) in Africa, the recurring famines in environmentally fragile areas and the breakdown in governmental infrastructure and the development of conflict in areas of high population, especially when

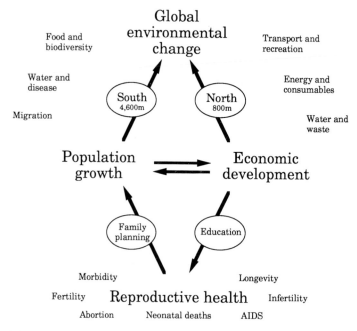

FIG. 1.   Schematic illustration of the principle relationships relating reproductive health to population growth, economic development and environmental change.

problems are aggravated by ethnic and religious differences. Those at the other extreme argue that mankind has the ingenuity to provide technical solutions to all possible constraints. There is, however, a reluctance within this school of thought to effectively address the issue of the planet's capacity to absorb insult (environmental destruction and the dumping of waste) or to acknowledge the enormous inequality of resource consumption between the developed countries of the North and the much less developed regions of the South*. Others combine the worst elements of both scenarios, and predict that some human activity, related to the inability to control advanced technology, will eventually precipitate an irreversible global disaster. The melt-down of Chernobyl in 1986 and the firing of the oil wells in Kuwait in 1991 were two recent events that were held to support that view.

Demographic update

Growth in global population is now without parallel in human history, both in magnitude and rate of change (Fig. 2). World population reached the 5000

---

*The terms 'North' and 'South' are used here to broadly separate the highly developed parts of the world (including all of Western Europe) from the far less developed regions of the equatorial belt.

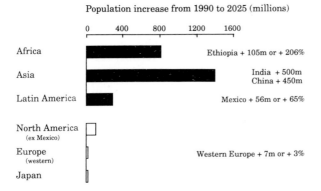

FIG. 2. Projected population increases (in millions) for three developed (open bars) and three developing (filled bars) regions for the period 1990 to 2025. Selected examples of these projected increases are given on the right, in millions of people and/or as a percentage of the existing population. Data adapted from World Bank (1992).

million milestone in 1989. A further 1300 million will be added to this figure before the year 2000. This projection is relatively firm, because it centres on a large and currently childless cohort of young people in the countries of the South and their potential for increased longevity. In broad terms, therefore, our planet has to accommodate within less than a decade an increase in population that equates to the current population of India, or China, or the entire developed world. Ninety-five per cent of this increase will occur in the already economically, environmentally and politically fragile regions of Africa, the Indian subcontinent, south-east China and Latin America. Projections beyond the year 2000 are subject to increasing error. Dramatic and rather unpredictable demographic changes have been observed after the introduction of family planning programmes, particularly when these have been backed by governmental pressures and incentives, as in China and Mexico. Under such circumstances, fertility rates have changed substantially within a decade. Likewise, it is difficult to estimate the impact of economic development (including education), the spread of the acquired immunodeficiency syndrome (AIDS) and the effectiveness of family planning programmes. If a rapid decline in fertility is established world-wide, akin to that recorded in Thailand and Mexico, global population could stabilize at about 10 billion (10 000 million) by the year 2085. More pessimistic projections, based on the changes recorded in Turkey and Sri Lanka, forecast stabilization at about 23 billion in the latter half of the twenty-second century (World Bank 1992).

Current demographic patterns exhibit enormous regional differences. Growth rates vary from plus 4% (Kenya) to minus 0.2% (Germany) per annum. Age structures vary from the youthful (over 50% under 15 years of age) to the aged (below 20% under 15 years of age) (Fig. 3). To a substantial degree, these differences represent a legacy from the time at which effective family planning policies were introduced. Developed countries, through education and health care programmes, introduced effective fertility regulation some 10 to 30 years ahead of many of the already highly populated countries of the South. Thus today we face a dramatic imbalance in the age structure in both developed and undeveloped parts of the world. In broad terms, the countries of the North have stable and ageing populations, with consumer-led economies which generate substantial pollution in relation to population size. The countries of the South differ markedly from those of the North, with an environmental impact resulting from the focus on the demand to supply food, water, fuel and shelter — and hard currency. This imbalance in the age structure will have serious consequences for at least two more decades.

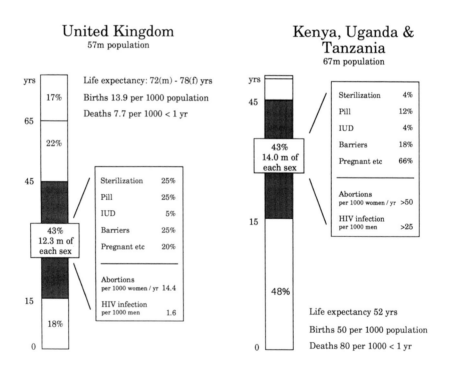

FIG. 3. Comparison of the population structure and reproductive dynamics of a developed (UK) and a developing region (Kenya, Uganda and Tanzania combined). The figures given for the developing region are best estimates from several sources. A breakdown of contraceptive practices among those aged 15–45 years of age (shaded region) is given (Pill, oral contraceptive pill; IUD, intrauterine device; m, millions).

Reproductive health and population growth

Women have the physiological capability, when provided with modern medical care, to produce 10 to 15 children over a reproductive life span of 30 years. This is greater than five times the fertility rate (2 to 3 children) required for population stability. The balance is established by interventional factors—contraception, sterilization, elective abortion and voluntary childlessness—and biological factors—lactational infertility (the absence of fertility whilst breast-feeding), genetic disease, spontaneous abortion, acquired infertility and neonatal death. In the past, lactational infertility and early death held population growth in check. Today, the key elements are sterilization, contraception and elective abortion.

Contraception and sterilization

Population growth could be regulated by currently available methods of contraception, but their acceptability remains marginal at a personal level. Most women in contemporary society require 25 years of protection against unwanted conception. Modern methods of contraception have failure rates of 1–10 pregnancies per 100 woman years, a combination of method and user failure. In mathematical terms, a failure rate of 10% per annum, when compounded over 30 years, would result in the birth of more children than required for population stability. It is estimated that 50% of pregnancies are not planned, and that half of these unplanned pregnancies were not wanted (Fathalla 1992). Only where family planning methods are rigorously applied, and coupled with wide availability of abortion and early sterilization, does one encounter fertility rates that fall below replacement level, as is now evident in some parts of Europe. Long-acting injectable contraceptives, steroid-releasing vaginal rings and low-level anti-gestagen therapy will shortly enlarge the choice of contraceptive methods. Contraceptive vaccines are under clinical trial (Griffin 1991), but are unlikely to have a practical impact within this decade.

Sterilization is the most widely used method of fertility regulation and has continued to increase markedly in both the North and the South (Church & Geller 1990). Sterilization is highly cost-effective because of its low failure rate and the number of years of contraceptive protection provided.

Reproductive morbidity

It is estimated that more than half of human oocytes fertilized fail to develop beyond the implantation stage. Much of this loss could relate to the expression of genetic defects that are not compatible with further development—in effect, there is a biological filter. Some 2–3% of babies are, however, born with a handicap caused by any one of 2500 inherited human diseases. Indeed, modern medical care is serving to increase the number of carriers of some diseases, and this could have significant consequences for health care in the early part of the next century.

More than 70% of the world's population now live in countries where abortion is legal, usually within a defined framework concerning stage of pregnancy and medical care. Some 480 million pregnancies will be terminated in the next eight years, with more than 30% involving the use of unsafe methods (Henshaw 1990). Rates are lowest in countries in which there are good education systems and access to modern family planning, as in most of Europe. Even so, one woman in three in Western Europe will have an elective termination of pregnancy at some time in her life. Some parts of the developed world report higher figures, including the USA, Japan and countries of Eastern Europe. The introduction of the medical termination of pregnancy through the use of anti-gestagens (such as the progesterone antagonist RU486) in combination with prostaglandins, could have a dramatic impact upon the abortion scene, and reduce maternal morbidity world-wide.

Maternal and neonatal deaths

It has been estimated that more women in the South will die during the 1990s from pregnancy-related causes than in any previous decade in human history. Women with septic abortions and the complications of pregnancy associated with large family size continue to fill gynaecological wards in the countries of the South, as was the case in Europe in the first half of this century. Of equal concern is the enormous level of childhood mortality. Some 25 000 children under the age of one year now die every day (9 million per annum), most from preventable causes. This is equivalent to the loss, for example, of one jumbo-jet, carrying 400 children, every 23 minutes — year in and year out. Indeed, the situation could deteriorate further if population growth continues unabated and if regional environmental changes destabilize the infrastructure such that the supply of health care, food and potable water is reduced.

Population growth and economic development

Perpetuation of a North–South divide

The economic division between the North and the South has arguably increased in the past 20 years (Table 1, Fig. 4). Technological developments in the North have continued to outpace economic development in the South, and resources — raw materials, finance and trained manpower — have continued to flow North. Indeed, the net flow of such resources under existing structures of international trade represents a form of economic rape, with greater store being placed on the provision of 'luxury wants' in the North than on 'subsistence needs' in the

TABLE 1 Division of world population by gross domestic product (GDP)

Economic classification	GDP (ECUs per capita)	Population (millions)	Population (%)
Low income	80–600	3058	58
Middle income	600–6000	1087	21
High income	7000–30000	816	15
Other economies (e.g. USSR, North Korea, Cuba)		321	6
Total		5282	100

Data from 1990 compiled from the World Bank's world development report (1992).
ECU, European Currency Unit.

South. The North has excess capacity in production of food and consumer goods; the South, by contrast, has an almost insatiable demand for such products but no hard currency with which to trade. Quite understandably, the South has sought to exchange its natural resources for short-term gains in hard currency (in the destruction of hardwood forests and open-cast extraction of minerals), generating major ecological changes that are largely irreversible. Likewise, agricultural activities have been substantially directed towards the production of goods for export (coffee, tea, sugar), rather than to the needs of the local people. Progressively, this traditional trading base has been supplemented by the supply to the North of commodities which are low priced because of the exploitation of cheap labour, with potentially serious consequences for the economies and health care systems of the developed countries. Ultimately, even

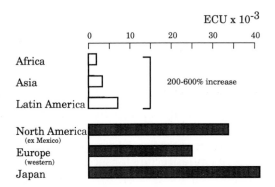

FIG. 4.   Projections of gross domestic product for the year 2025 for three developed (shaded) and three developing (unshaded) regions. The percentage increases projected for the developing regions relate to those existing in 1990. Data are adapted from World Bank (1992).

modest economic development in the South may generate a feedback loop in which cheap exports reduce the economic well-being of the North, thereby reducing the availability of aid.

Economic, environmental, political and intellectual migration

Pressure from people in the South to migrate to areas where they may achieve a higher standard of living (economic migration) is increasing (United Nations Population Fund 1992). Language, the perception of opportunity and declining living standards are some of the more important determinants governing the flow of people. The impact on the North will be progressive but acute if the rate exceeds its capacity for absorption. Economic migration continues to gain momentum, especially where there is an abrupt interface between rich and poor, as between Eastern and Western Europe, Mexico and the USA, and South-east Asia and Australia; such movement threatens to become increasingly difficult to police.

'Environmental' migration has hitherto been slow and progressive and its impact has largely passed unnoticed. The gradual change of North Africa from savannah to desert over the past 10 000 years, largely as a consequence of climatic change, provides one example. In the future, however, crop failure due to recurring droughts in environmentally fragile areas, amplified by poor agricultural practices, could precipitate massive movements of people within a span of a few years. The current movement of people into South Africa from the adjacent regions of Zimbabwe, Botswana and Mozambique, in response to an eight-year drought, could provide a foretaste of larger migrations to come.

Conflict, fuelled by population pressures, ethnic differences and their associated politics, appears to be increasing in many regions, with the potential to impose disaster on the displaced people and the adjacent countries in which they are forced to take refuge. Examples of these problems are evident in many regions, including the Middle East and even Europe.

The consequences of migration for Europe, the USA and Australia could be very serious, placing pressure on social, housing and health care sectors, generating ethnic conflict, and altering patterns of disease and employment. There is another serious problem for which the North might be held responsible. International aid for economic development has in part financed a 'brain drain', because a significant proportion of the people from the South who have received training overseas have not returned to support the development of their home countries. Thus, some regions of the South have lost the trained people required to establish and support an infrastructure for successful economic development.

Acquired immunodeficiency syndrome

The impact of AIDS on individual countries will vary dramatically over the next decade. It is estimated that 2–4 million people will have died from AIDS-related infections by the year 2000, and a further 100–120 million will have become infected with HIV 1 and 2. These figures, although serious, are modest when set against the 800 million projected increase in population. However, the continued spread of HIV will place an almost insurmountable burden on some regions of the South, particularly those where infection is projected to rise steeply.

The level of AIDS in parts of sub-Saharan Africa has reached tragic proportions. It is estimated that some 6 million men and women of reproductive age are now infected with HIV in this region. This represents one person in 40, though figures are five-fold higher in some urban areas. AIDS-related infections, notably tuberculosis, are likely to kill 15–20% of the work-force before the year 2000. In Tanzania, for example, agricultural production has already fallen by 10–20% because of AIDS-related morbidity within the work-force. This, in turn, may result in a larger proportion of children, including those orphaned by AIDS, being drafted into the work-force, which would reduce standards of education and, in the longer term, would adversely affect economic development. The second effect will centre on medical care. Eighty per cent of the case load in some hospitals already involves AIDS-related infections, yet the full burden of the disease has still to express itself.

The speed with which HIV infection may spread in other parts of the world is also a matter of considerable concern (Potts et al 1991). AIDS could seriously affect India and South-east Asia in the next decade, because of the high population density, the pattern of sexual activity, the low social standing of women and the high incidence of sexually transmitted infections — an indicator for the spread of HIV. Estimates of the number of infected individuals in India and Thailand, as of 1992, has recently been increased from half to one million. Within some of the large cities, such as Bombay and Madras, the high level of infection amongst prostitutes, which now exceeds 20%, could cause the infection to spread rapidly within the migrant Indian workforce. Likewise, some of the major cities of the South could serve to disseminate HIV to Europe and elsewhere.

The cost of family planning

About 350 million people in the countries of the South have access to methods of family planning. By the year 2000, this figure has to rise to 600 million if median population projections are to hold, because of the young age structure in these regions (Lande & Geller 1991). The medical cost of providing family planning in the South will amount to about 9000 million ECU per annum by the year 2000 — a substantial figure in relation to the financial resource base of these regions. Currently, family planning costs in the South are met by national governments (63%), through donations from other countries (20%), and by users (13%). It has been estimated that only 1% of total developmental assistance goes towards the support of family planning programmes. Clearly, a small shift in developmental aid in favour of fertility regulation could transform the entire picture.

Pharmaceutical developments

Much of the South does not have the wealth to finance what the North would regard as the minimum standards of health care. Unfortunately, the protectionist policies and corporate profit motives of the international pharmaceutical companies are structured solely to the needs of the most advanced countries. Investment will continue to focus on drugs intended for use in the developed world, with particular attention being given to the treatment of neurological and cardiovascular problems. Humanitarian gestures to help restrict population growth in the developing world have no significant place on the corporate balance sheet.

The problems of finance and the pharmaceutical industry assume a critical dimension in the area of contraceptive products. The problem is two-fold. Contraceptive products are designed for use by well people over many years. Modern oral contraceptives have been subject to perhaps more research and post-market surveillance than any other group of pharmaceutical products. When used with appropriate medical supervision, the modern generation of oral contraceptives has the potential to increase the average lifespan of the woman—that is, positive effects outweigh negative effects (Fortney et al 1986). However, negative effects, largely related to breast cancer and cardiovascular embolism, have been exaggerated. Neither the media nor the legal profession has an adequate appreciation of the concept of relative risk. Thus, pharmaceutical companies have had to face massive litigation costs. The second difficulty is that contraception and abortion have become inextricably intertwined, leading to the black-listing of pharmaceutical companies by pro-life organizations. Likewise, many political figures have sought to side-step the issue for fear of losing votes. Thus, the question of population was not allowed on the agenda at the 'Earth Summit' held in Rio de Janeiro in 1992. As a consequence of such attitudes, pharmaceutical companies in the North have withdrawn their investment in contraceptive research and development on a massive scale in favour of a safer investment elsewhere, yet the need (and opportunity) for the development of new strategies for the regulation of human fertility has never been greater. A significant pharmaceutical capability is developing in China, Mexico and South-east Asia, with a significant focus on the regulation of fertility. This could resolve some of the problems of the South, if the effort is adequately supported by scientific input from the North, but equally, it could seed a new drugs war. Already, centres in South-east Asia and elsewhere are manufacturing medications that could furnish a black market in which abortifacients and AIDS-related drugs could feature.

Population growth and environmental change

Food, land use and biodiversity

Rising income, as a feature of economic development, has a major impact on the environment. In the South, an increase in the level of income would, in all probability, stimulate a large increase in demand for basic foods, especially when the prevailing conditions are close to starvation. As income rises further, demand would switch to more protein-rich foods, processed foods and consumer goods. With a high elasticity in demand for food, increases in per capita income can generate a greater increased demand on the land resource than that associated with the housing of the people. Indeed, the increased demand on agricultural production becomes monumental when the two factors—an increasing population base and an increasing per capita income—are combined.

The proportion of the world's people that require an increased food intake is already substantial. Some 340 million people currently have a calorie intake that is below 80% of the FAO/WHO calculated demand, that is, the level required to prevent stunted growth. A staggering 730 million people fall below the 90% figure, that is, the requirement for an active working life, and 40% of these are children below the age of 10 years. The effects of poverty tend to differ markedly between rural and urban areas in the South. Rural areas are often environmentally fragile and subject to degradation. Work and food availability have a seasonal and unpredictable character. Those living under such rural conditions of poverty have traditionally invested in children in the hope that a few will survive to generate income and support for aged parents. For them, a small family is a high risk strategy. An increase in the number of people in the South of a further 800 million in the next eight years will, in all likelihood, increase the number of children living on an inadequate level of nutrition. Subsistence agriculture will increase in an attempt to meet food requirements, in the absence of massive financial investment. As a consequence, land will be degraded at a faster rate, productivity will fall and erosion will be increased. It is estimated that 11% of the vegetated land surface of the planet, some 1200 million hectares, has undergone serious degeneration in the past 40 years (Oldeman et al 1990, cited by the World Bank 1992). Bad irrigation practices have added to the problem by increasing salination, in Egypt and areas surrounding the Aral Sea, for example. Collectively, these changes have already led to a decrease in crop yields in areas where population is now set to expand markedly. The impact of agriculture will be compounded where the opportunity exists to pillage natural resources in the quest for hard currency. Most notably, this will result in the destruction of the hardwood forests of the equatorial belt. Returning these forests to their current structure on any significant scale, once destroyed, will not be possible in the foreseeable future.

In the developed countries of the North, by contrast, elasticity in demand for food according to income is close to zero, but is high (and labile) for recreation, housing, energy and consumer goods. This creates a demand for space unrelated to food production, and that is environmentally destructive when related to the increased demand for water, energy, transport, recreation and waste disposal. Indeed, an ever-increasing level of food production per unit area of agricultural land, coupled with a saturated demand for food from a static and ageing population, has already resulted in over-production, but at a unit cost that, in convertible currency, prohibits export to those regions of the world that are in need of food. The challenge, in the North, now centres on the redevelopment of land for purposes other than food production, with the emphasis on environmental benefit.

Water and disease

Water is a key resource, required for personal consumption, hygiene and health care, and food production. Fresh water is not in short supply on a global basis, but most of it is in the wrong place. The population growth in the South will generate a substantial increase in the need for water. However, this demand will increase by a far greater factor if one adds in the demand that is associated with economic development (Table 2).

The provision of good quality drinkable water at source and sound methods of sanitation forms the most effective investment that can be made to improve both health (reducing diarrhoeal diseases, roundworms and schistosomiasis) and economic productivity—a fact known to the developed world for at least a century. Sanitation is grossly inadequate in the urban slums of all cities of the South, with contamination of surface water supplies by human excrement threatening the outbreak of water-borne diseases. Several of the largest cities of the world, including Mexico City and Shanghai, now have to import water from increasingly large distances, because of a combination of raised demand, pollution of local supplies and climatic changes.

Serious problems also exist in the South in the disposal of industrial by-products, some imported from the North in exchange for hard currency. The open discharge of toxic waste is made more serious by the fact that industrial developments are often associated with high population densities. In the longer term, the discharge of such pollutants as lead, cadmium and mercury could largely irreversibly contaminate ground aquifers, which could further limit the availability of local water supplies. The allocation of water for personal consumption (and hygiene) and food production continues to present the South with a problem of considerable political magnitude. Should one supply the people with potable water to maintain health (and thus economic viability), or support the production of food for export? Peru has recently paid the cost of supporting the latter stance. The spread of cholera not only had a substantial effect on human health but is estimated to have also reduced food exports by more than the equivalent of 1000 million ECU, because unclean water was used

TABLE 2   Annual renewable water supplies and water use for selected regions in cubic metres per capita per annum

	Total water used	Domestic	Industry	Agriculture
Low income[a]	391	16	20	355
Ethiopia	48	5		
Nigeria	44	14		
India	612	18		
Middle income	450	59	81	311
Mexico	901	54		
Hungary	502	45		
High income	1170	164	550	456
France	728	116		
Japan	923	157		
USA	2162	259		

[a]Income brackets are as defined in Table 1.
Data compiled from the World Bank's world development report (1992).

in food processing. The question of food production is further complicated by the fact that most intensive forms of production focus on products for the developed world rather than home consumption.

Air pollution

The people of the South, now exceeding 4000 million, generate enormous pollution in areas of high population density, but that pollution tends to be local in nature. Air quality in most (if not all) large centres of population in the South is extremely poor. It is estimated that some 1300 million people (25% of the world's population) are exposed for most of the year to a level of suspended particulate matter in the atmosphere that exceeds WHO guidelines. Even the least polluted cities in the South have poorer air quality than the most polluted cities in the North. This must increase morbidity from lung diseases, reduce productivity through ill health, and shorten the lifespan.

The 800 million people of the developed world generate a much higher per capita level of pollution, primarily from manufacturing industry, energy production and personal transport. The atmospheric emissions that typify such pollution have local, regional and potentially global consequences. Lead and ozone are two of the most toxic agents in areas of high traffic density. Sulphur, emitted from power stations in particular, causes production of acid rain, the effects of which can be felt several hundred kilometres downwind. Other gases, notably carbon dioxide, methane and the chlorinated fluorocarbons have world-wide effects, with the potential to influence climate and destroy the protective high level ozone layer. The extent of most of these emissions is related to both population size and level of economic development, given that the raw materials on which these emissions depend are not rate-limiting at present. Errors of the past will continue to introduce a hitherto unbudgeted for element into the economic equation. The cost of decommissioning the nuclear and heavy industries of Eastern Europe and the former USSR provides one example. Public pressure for the generation of 'clean' power, manufacturing, transport and waste disposal systems will grow, but is unlikely to be evenly spread throughout the developed world. This stands to place those countries who have taken an environmental lead at an economic disadvantage to others.

Feedback regulators

Environmental infertility

There is no evidence that infertility, caused by pathological or environmental factors, is rising at a rate that could alter population growth. However, there is increasing evidence that the human reproductive system, notably of the male, could be more prone to disruption by environmental toxicants than has hitherto been appreciated (Skakkebaek et al 1991). High on the list of toxicants are pollutants from heavy industry, and even products to be found in the domestic environment. Plasticizers, quick-drying agents and dietary additives have all been implicated, at least with high exposure. It is far more difficult to determine whether low level exposure over many years has similar consequences. One recent evaluation suggests that the human sperm count may have fallen by as much as 50% over the past 50 years, but such figures could be confounded by sampling bias, technological advances and changes in sexual practices (Carlsen et al 1992). In terms of fertility and population growth, however, it hardly matters whether a man produces 1000 or 500 sperm a second. The incidence of testicular cancer has also been reported to have risen three-fold (see Carlsen et al 1992 for references). This is more interesting, and is unlikely to be confounded by observational bias. Of equal concern is the possibility that prenatal and postnatal exposure to chemicals may have harmful effects on sexual development which may not be revealed for two decades. Similar concern now centres on the connections between prenatal development and the risk of cardiovascular disease and cancer in adult life.

The structure of the work-force

Voluntary childlessness has been rising markedly in the developed parts of the world, to the point that the birth rate in some European countries is now below replacement level. If this continues, there will be a substantial alteration in the age structure of these countries, and in the longer term an even greater imbalance with the rest of the world will result. It is difficult to ascertain whether this change relates to economic well-being or a direct wish not to bring more children into a world where conditions threaten to deteriorate. Likewise, the increasing tendency of women in some developed countries to follow professional careers has altered family structures. In the UK, the mean age at first delivery has advanced to 27 years. This is almost a decade beyond the peak of a woman's reproductive potential, and is beginning to encroach upon the age at which the chance of medical problems associated with pregnancy rises sharply. Certainly, by this age, many women experience acquired infertility, often the result of the additive effects of subclinical infections. Assisted conception procedures have come to the aid of these people. However, these procedures have dramatically increased the number of multiple births in the UK in the past seven years, placing a substantial new charge on health resources. In addition, the well-being of these children could be compromised; on average, triplets have a lower IQ (intelligence quotient) than singletons.

Education of women and provision of family planning

The provision of education and family planning for everyone in the South has the power to halt population growth and facilitate sustainable economic development. Poverty and illiteracy are invariably associated with large family size and poor health; this is true in both developed and undeveloped countries. Furthermore, where opportunities are marginal, more importance has to be attached to the education of women than men. Educating a mother educates a family, in the context of both her children and her use of family planning. In the absence of that education, world population will continue to expand, economic development will remain on hold, and environmental degeneration (with possible global consequences) will continue to gather pace.

References

Carlsen E, Giwercman A, Keiding N, Skakkebaek NE 1992 Evidence for decreasing quality of semen during past 50 years. Br Med J 305:609–613
Church CA, Geller JS 1990 Voluntary female sterilization: number one and growing. Johns Hopkins University Population Programme, Baltimore, MD (Popul Rep Ser C 10)
Fathalla MF 1992 Reproductive health in the world: two decades of progress and the challenge ahead. In: Khanna J, Van Look PFA, Griffin PD (eds) Reproductive health: a key to a brighter future. Biennial report 1990–1991 special programme of research, development and research training in human reproduction. WHO, Geneva, p 3–31
Fortney JA, Harper JM, Potts M 1986 Oral contraceptives and life expectancy. Stud Fam Plann 17:117–125
Griffin PD 1991 The WHO Task Force on vaccines for fertility regulation. Its formation, objectives and research activities. Hum Reprod 6:166–172

Henshaw SK 1990 Induced abortion: a world review 1990. Fam Plann Perspect 22:76–89
Lande RE, Geller JS 1991 Paying for family planning. Johns Hopkins University
  Population Programme, Baltimore, MD (Popul Rep Ser J 39)
Potts M, Anderson R, Boily M-C 1991 Slowing the spread of human immunodeficiency
  virus in developing countries. Lancet 338:608–613
Skakkebaek NE, Negro-Vilar A, Michal F, Fathalla MF 1991 Impact of the environment
  on reproductive health. Report and recommendations of a WHO international
  workshop. Dan Med Bull 38:425–426
United Nations Population Fund (UNFPA) 1992 State of world population 1992. Nuffield
  Press, Oxford
World Bank 1992 World development report 1992: development and the environment.
  Oxford University Press, New York

DISCUSSION

Bradley: There is now less of a dichotomy between countries in terms of wealth
than there used to be. There are many countries which are neither rich nor poor,
or are changing between categories.

Has anyone obtained the patent for RU486 from the firm concerned (Roussel)
since they don't want to sell the drug? If this could be purchased by IPPF
(International Planned Parenthood Foundation), for example, to get it into the
public arena, the world might save itself an enormous amount of money at a
later date.

Lincoln: The WHO Human Reproduction Programme took an important
lead here, and a gamble in some respects. It should be noted that they are a
'special programme', and, as such, are funded separately from the WHO
organization itself. Roussel allowed them to distribute RU486 for research. Also,
the Chinese decided to develop anti-gestagens themselves, and their chemists
have discovered ways of making anti-gestagens with a much improved yield.

There are a range of other anti-gestagens under development by other
companies which are every bit as effective as and perhaps more effective for
some purposes than RU486. Unfortunately, these companies have been reluctant
to take these compounds into clinical trials. They are developing them, but
perhaps only to keep on the shelf for a rainy day, just in case some of the existing
technologies get taken out of the system.

Kleinjans: You talked about the decline there has been in male fertility. I
once heard a talk given by a urologist from Brussels, who analysed urine from
sperm donors and found a relationship between subfertility and compounds
such as glycol ethers, which are found in paints. The main message was that
those males suffering from infertility were not painters by trade, but were
probably encountering these chemicals while painting their houses in their free
time. Do you know which chemicals are involved in this decrease in male
fertility?

Lincoln: We have been looking at the question of spermatogenesis for many years, using a variety of chemical agents to selectively destroy specific cell types. We have discovered that some cell types within the testis—the Leydig cells and the secondary spermatocytes—are very sensitive to damage by toxicants. Some of the chemicals we tested originally came under suspicion after reports of infertility after industrial exposure. We know that the solvents that were used in some of the quick-drying paints and the plasticizers that were used in food wrappings were dangerous in this respect. We suspect that there may be thousands of chemicals out there which have the potential to selectively target the testis.

We don't fully know the cause of this long-term reduction in sperm count. We do know that sperm production relates to the number of so-called nurse cells or Sertoli cells in the testes, and that the number of Sertoli cells is laid down during late fetal life and the first few weeks of postnatal life. We also know that hormones given to animals during this critical period can change Sertoli cell numbers, and can also generate the abnormalities of sexual differentiation that we are now seeing in the human population. We suspect that the important time in life is therefore the late fetal, early neonatal stage. High level exposure in an industrial context and low level chronic exposure are two different issues, of course.

Lake: As has been described for effects on sperm production, there can be a latency ranging from 20 to 30 years before the effect of an environmental change, particularly a chemical one, becomes evident. Similarly, some of the respiratory disorders of today appear to reflect exposures 20–30 years ago. We are faced with an intrinsic problem, the difficulty of making an experimental or analytical link between an environmental change and its consequences for human health. Although we might address the matter with all urgency at this meeting, we will still be left conjecturing about the course to take, because we don't have enough evidence from 20–30 years ago. Perhaps modelling is the most promising approach, because you can build that sort of lag into models and you can build imprecision into models.

James: What one needs to do therefore is construct a biologically sensible sequence of events using intermediate markers of the processes. Western policy-makers have to make appropriate judgements by integrating information, yet we are assailed constantly by our fellow scientists for being sloppy in the way in which we go about the linkage and integration which are fundamental to the process of making policies and decisions about any complex biological phenomenon. You have to integrate and make judgements, and you can be wrong.

Elliott: In cancer epidemiology building in lag periods is a well-established technique. In occupational cohort studies, for example, one has to follow up workers over many years, and in some industries the data exist which enable one to build in lag periods between exposure and the first disease outcomes, discounting diseases that occur within the first 5, 10 or 15 years. For example, in a recent study we conducted on health effects around incinerators of waste solvents and oils we built in a lag period—events occurring within 5–10 years of the beginning of the site's operations (or first registration) were not included in the analysis. If you want to look at the effects of current exposure, you can't use cancer as an endpoint. You could look at respiratory events as an endpoint for current air pollution, but if you wanted to study whether biomarkers predict cancer you would have to wait 30 years.

Lake: Or use a surrogate marker for the eventual cancer, as Professor Kleinjans described.

Elliott: You need validation that those biomarkers actually relate to the disease. Ultimately, we need to count health events, disease events, in individuals and populations.

Kleinjans: There are study designs and methodologies to cope, to a certain extent, with analytical problems, but latency periods occur all over the place. The time taken for migration of pollutants through the soil is another example of a latency period. Dutch environmental policy-making is designed first to manage the sources of environmental deterioration and second to manage the effects until the source is under control. Remedial measures are required until the problem is eliminated at source. That takes about 20–30 years, so there are two different types of latency period to be considered.

Bobrow: I think we can be optimistic that the increase in our knowledge of the molecular basis of cancer is now fast enough that within a few years we will be able to define intermediate, short-term effects at a molecular level, thus eliminating this problem of the latency period. If that doesn't come to pass, I agree that you just have to make your best guess and get on with it, because you can't defer policy decisions for 30 years waiting for validation. However, there is an important corollary to this. Having taken that operational decision, you should not then stop work aimed at proving or disproving the validity of the decision, as has happened occasionally.

General discussion

Avnimelech: Although the issue of nitrates is not on the agenda of this meeting, I thought it would be an appropriate place to raise it, particularly as we have discussed interactions between science and policy-making. European Community standards for nitrates in drinking water are in the process of being reduced. Huge amounts of money have been and will continue to be spent on treatment of water to reduce the amount of nitrates. What is the epidemiological or medical basis for this? There is of course the blue baby disease (methaemoglobinaemia) to worry about, but this is a rare disease and there is no problem for the baby if he is breast fed, if he is given ready-made formula or if he is getting vitamin C. As far as I know, there is no conclusive evidence that nitrate at a concentration of up to even 90 mg/l has any adverse effect on adults. Moreover, the amount of nitrates that we take in food is much greater than that we get from water. There is a lot of nitrate in spinach and lettuce and other vegetables. Do we really need to limit nitrates in drinking water to the concentrations planned? The only available technology for removing nitrates requires methanol to be added to the water, so we have to ensure that all the methanol is removed and that there are no by-products.

Kleinjans: We have been studying the effects on health of nitrate contamination of drinking water for several years now. The question is whether the current standards for nitrates in drinking water are adequate or not. Generally, the EC, WHO and national standards will protect against induction of methaemoglobinaemia in children, except for one high risk group, very young bottle-fed infants, with a weight of about 5 kg, who consume about two litres of water per day. There is no real safety margin for that specific group, but the current standard would prevent any harmful effects for that high risk group also.

The second aspect is the induction of stomach cancer. This is being discussed by WHO. Nitrates from water might form nitroso compounds in the stomach. The evidence is controversial; there are two groups of epidemiologists, in Italy and in Denmark, who have found an association between the occurrence of stomach cancer and nitrates in water at concentrations around 25 mg/l, a concentration lower than the current drinking water nitrate standard, but various groups in the UK have found that the concentration needs to be at least 100 mg/l for a reliable association. We have studied the genetic effects in various populations in The Netherlands exposed to different concentrations of nitrate; we found no correlation between nitrate concentrations and the occurrence of cytogenetic dysfunctions.

Nitrates may also be involved in the induction of thyroid dysfunction. Nitrate acts as a competitive inhibitor of iodine uptake by the thyroid. With sufficiently high levels of nitrate, thyroid volume increases and, over the longer term, a goitre may develop. This effect was established for nitrate concentrations in drinking water above the current standard of 50 mg/l, but our database was not large enough for there to be conclusive evidence for an association at concentrations below that standard. Nitrate contamination of surface and ground-water is still increasing predictably, to concentrations that will exceed the current nitrate standard for drinking water by the year 2010.

de Haan: You suggested that the current nitrate standard, 50 mg/l provides sufficient protection against methaemoglobinaemia. There is a movement in the EC to lower the standard to 25 mg/l. To satisfy this standard many countries will have to spend millions. If there is no scientific reason for this, how can we influence the people setting such European standards?

Kleinjans: We have to present the data to them. Also, we have to face the fact that even the current standard of 50 mg/l will in the near future be difficult to meet with current purification technology. The cost of drinking water will go up anyway. A single woman with two children living on a limited budget who has to pay more for drinking water will have less money to buy good food.

de Haan: Already in The Netherlands there are a number of drinking water wells which do not meet the current nitrate standard. At Wageningen Agricultural University a method was developed for removal of nitrate from drinking water, a combination of resin treatment and a biological treatment, but there is a tremendous amount of money involved.

There is also a drinking water standard for potassium in the EC, a maximum concentration of 40 mg/l, and nobody knows what the basis is for this standard. This is a political development I cannot understand; we will run into problems if this potassium standard is taken seriously.

Klein: How is the acceptable level of nitrate scientifically established? The concentration of nitrate in drinking water has been increasing over the last two or three decades, and we are now in a peak situation. From initially low levels, we integrate over all values lower than the presently accepted 50 mg/l. This doesn't mean that at a steady-state of 50 mg/l everything will be fine. This is why people want to reduce the standard to an average value between 0 and today's 50.

The problem of pesticides in drinking water is well known. The limiting value is purely legislative and protective in nature, and has no relevance to any toxic effects in humans. The legislation means that pesticide manufacturers have extreme difficulty introducing any new pesticide which is soluble in water.

de Haan: The problem is the purism within environmental protection policy. Policy-makers simply say that xenobiotics do not belong in the water system— full stop. The difficulty will increase as our ability to detect compounds at lower concentrations increases. If interpreted to the letter, the policy is effectively a complete ban on pesticide use. I would of course be in favour of that if we could come up with alternative controls for plant disease. In a healthy soil system you have to expect these compounds to be present, because you want soil that supports root systems and earthworms etc., and hence a short shunt between the top layer of the soil and the ground-water is unavoidable. All these compounds will be found in ground-water at low concentration.

Zehnder: You are correct, but when I drink water, I would like to drink water and not a mixture of pesticides. Norms and standards catalyse developments in science and technology. Standards should not be used to sharply separate black and white; rather, they should indicate the extent of a grey zone. Standards should be based on sound scientific information; they should not just fall out of the sky.

Bradley: It's fine to decide on a particular standard, and make it rather rigid to encourage developments, but there are costs to consider. There are major problems in the world which are more overt but get neglected. As I found when I was involved in the establishment of the last but one set of water standards at the WHO, there are two quite different philosophies. Some people, including some of those from less developed countries where the quality of drinking water was abysmal by any standards, wanted to set much higher standards than most of us did. They said that they wanted a target at which to aim. That is one philosophy, to set the target up in the sky. The pragmatist's philosophy is to bring the targets down, to set them at a level which is realistic for more people, so that the standards could be enforced even at the village level. These philosophies need to be discussed overtly and not covertly, otherwise people simply fight.

Sutherst: I have difficulty with the concept of having a tolerance level defined as acceptable because it's in some way convenient. If concentrations of nitrates and phosphates and pesticides in water are higher than they used to be, there's something wrong. Rather than establishing or raising permitted levels to accommodate present practices, why not question present practices? It seems to me that people are pouring far too much fertilizer onto plants, that too much nitrate is coming out of the cows onto the pastures because there are too many cattle, and that there is too much pesticide usage, because when nitrogen is put onto crops the resistance of the crops to attack by pests is reduced. These standards for water shouldn't be limited to measures of potential effects on human health only, but should also be indicators of the general state of the environment. I would say that, if anything, the nitrate standard should be reduced to 25 mg/l, to force changes in agriculture which will in fact be highly beneficial to the health of humans and the environment in general.

Edwards: Potassium in drinking water is obviously extremely harm*less*, and it seems to me serious that we find ourselves in circumstances under which a potassium standard can find its way into legislation. We should be worrying about the Third World. If it's possible for such an absurd piece of legislation to be passed by a group representing a higher standard of living this side of the Atlantic, this is extremely dangerous. As soon as somebody tries to construct something which is urgently needed in a developing country, such as a desalination plant in Ethiopia, there will be an immediate demand that it should conform to the standards set in Europe. Standards with no relevance to scientific or health matters are bound to reduce the ability to provide water and other things in places where they are needed. Such an irrational policy is not a trivial matter affecting only Europe.

The same is happening in radiation protection. The permitted radiation level is so low that after the Chernobyl accident deer were shot in Lapland, disrupting the Lappish economy. At the moment lambs in Wales which cannot be sold because they contain too many bequerels are being left to grow into sheep which contain more bequerels in total but less bequerels per kg so they pass EC regulations and can be eaten. There are 45 nuclear power stations in France. If one of them blows up when there's a south wind blowing, the entire agricultural produce of the UK would become illegal to eat. There are practical issues involved in absurd legislation being pushed through in the absence of adequate scientific scrutiny.

Sutherst: The use of energy efficient light bulbs which produce ultraviolet light is being encouraged to reduce energy consumption. Are there any consequences to health? Also, irradiation sterilization of fruit to reduce pesticide use is being proposed. Are there any likely effects on public health?

Bobrow: There are some lights of the quartz-halogen variety which produce a significant amount of UVB. Only in an occupational setting is a person likely to accumulate a dose of UVB higher than that received by a northern European on his average fortnight's holiday in the Mediterranean. A piece of glass in front of the light prevents UVB exposure.

I have been told that the low LET radiation used in irradiation of foodstuffs produces no residual radionuclides, and therefore any effect is transient.

James: I sat on the British committee that examined the issue of food irradiation. The problem that arose with the Indian government's views depended on the nature of the data from one of the major laboratories where there were some much disputed and odd changes in the blood of malnourished children fed irradiated wheat. For the irradiation of fruit to deal with pests you can use a very low dose. If you are trying to arrest ripening of fruit or vegetables you are still dealing with very low doses; the exposure depends on the nature of the material. Any particulate matter, for example, associated with the fruit, or a food commodity containing fat, presents a different problem because there is some residual radiation which decays rapidly over about 24–48 hours. The problem then is one of the logistics of irradiation and the transfer of food into the food chain. The difficulty is again the public perception of risk in relation to this unknown problem, which they think is ill-defined and controlled by scientists and industry whom they consider to be intrinsically unreliable. They believe these processes can affect health in some mysterious way over which they have no control. There is no real problem in scientific terms, but there is a tremendous problem in public perception terms. The presentation of this issue to the public was in fact grossly mishandled.

Global epidemiology

Paul Elliott

Environmental Epidemiology Unit, Department of Public Health and Policy, London School of Hygiene and Tropical Medicine, Keppel Street, Gower Street, London WC1E 7HT, UK

Abstract. Epidemiology is the study of the distribution and determinants of health and disease in human populations. Epidemiology on a global scale is severely constrained by the lack of data. In many countries, there are no comprehensive data on mortality or basic demographic data. Where data are available, findings on the relationship of environment to health across countries need to be interpreted with caution. For example, there is well-known variation in standards of medical practice and diagnosis, and in certification and coding, but there are also large differences in diet, the social environment and lifestyle—all of which strongly predict disease incidence. Inappropriate inference concerning aetiology made from such broad-scale studies may result in what has been termed the 'ecological fallacy'. A complementary approach is to collect and analyse data in standardized fashion as part of international collaborative studies. These can offer some important advantages over the more conventional single-centre design. Recent advances have meant that studies of environment and health can now—in some countries—be carried out using routine data at the small area level. Although problems of interpretation remain, they are generally less severe than in broad-scale studies. Examples of this approach are given.

1993 Environmental change and human health. Wiley, Chichester (Ciba Foundation Symposium 175) p 219–233

Epidemiology is the study of the distribution and determinants of health and disease in human populations. It is therefore crucial to any discussion of the effects of environmental change on the health of human populations. The purpose of this paper is to present some of the epidemiological approaches to the study of environment and health within an international or 'global' context.

In this symposium, environmental change is being considered along a spectrum, which ranges on the one hand from the global effects of the greenhouse gases and the possibility of climatic change, and on the other to the study of exposure to environmental contaminants at the cellular level. Epidemiological enquiry similarly may reflect this spectrum, contrasting what might be termed 'macro-' versus 'micro-' epidemiology.

The spectrum is summarized in Fig. 1. At the *cellular* level, studies in humans include the use of biological markers, as discussed elsewhere in this volume

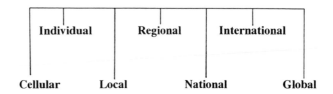

FIG. 1.   Levels of enquiry into the relation between environment and health.

(Kleinjans 1993). From an epidemiological perspective, long-term follow-up of large cohorts variously exposed to suspected environmental carcinogens has been proposed, to see whether biological markers (retrospectively analysed, for example, from banks of stored blood) predict disease outcome (Doll & Peto 1981). At the *individual* level, classical case-control or longitudinal (cohort) studies are usually considered necessary to test epidemiological hypotheses that may have arisen from broader scale (i.e. ecological) analyses. The methods are extensively described elsewhere (Breslow & Day 1980, 1987) and are not considered further here. At *regional* and *national* levels, many studies have been published describing and analysing geographical variations in disease rates (Gardner et al 1983, Kemp et al 1985, Teppo et al 1980, Carstensen & Jensen 1986) but, as Barker (1981) has remarked, 'maps of disease compel speculation about aetiology, but only rarely has such speculation by itself led directly to the discovery of causes'. Further discussion on this point can be found in English (1992).

Here, examples of the epidemiological approach at the remaining three levels of enquiry shown in Fig. 1 (*global*, *international* and *local*) are presented, together with some discussion of their strengths and weaknesses. First, it is necessary to define the terms 'environment' and 'environmental change'. The definitions need to go beyond the man-made and physical environment, to encompass the political and social environment, and include changes in individual exposure to 'lifestyle' factors such as diet and smoking behaviour. This broader definition of the environment is adopted here. Only then can we begin to understand some of the complex interactions that determine the risk of disease in individuals, and the occurrence of disease in populations.

Studies of changes in the environment at the global level

A classical approach in epidemiology is to study the geographical variability in disease rates across areas with contrasting levels of exposure to some environmental contaminant. Such studies may initiate or strengthen hypotheses as to causality which are often then tested in individual-based studies (case-control or cohort). When the effects of global environmental change are studied, however, this opportunity for further investigation at the individual level is lost;

by definition, whole populations are exposed to similar levels of the relevant environmental agent (for example, increased levels of UV light secondary to thinning of the stratospheric ozone layer).

One important epidemiological response to environmental change at the global level would comprise international comparison of national disease rates, and the study of disease trends over time. A major difficulty is that epidemiology on such a global scale is severely constrained by a lack of data. The information required for even simple descriptive studies includes valid measures of disease frequency (e.g. mortality data) in defined population groups and basic demographic data (age and sex distributions). Such data do not exist in many countries.

In countries where demographic and disease data (mainly mortality, and in some countries cancer incidence data) are readily available, they have proved useful in international comparative studies and studies of time trends, despite problems of interpretation (Anonymous 1992, Doll 1991, Lopez 1992, Muir et al 1987). These problems include international differences and changes over time in the diagnosis of cause of death or cancer; changes in certification practices and coding; incomplete and duplicate registrations of cancers (all-cause mortality data are essentially complete); and international differences in rates of diagnostic confirmation by autopsy or histological examination (Anonymous 1992, Kelson & Farebrother 1987, Lopez 1992, Mackenbach et al 1987, Percy & Muir 1989, Swerdlow 1986).

Despite these difficulties, there do appear to have been real changes in the incidence of cancer, for example, increases (or, more recently, a levelling-off) of lung cancer rates in Western industrialized populations, decreases in stomach cancer rates world-wide, and in some countries increases in melanoma and prostate cancer (Anomymous 1992, Muir et al 1987). Further problems of interpretation arise when considering the cause of such changes. Most diseases are multifactorial in origin, the causes being a complex mix of genetic and environmental influences. For the major killing diseases of the developed world—cardiovascular disease and cancer—environmental factors are of overwhelming importance, in particular diet and cigarette smoking. It has been estimated that some 20% of premature mortality (under 69 years) in developed countries is currently caused by cigarette smoking, and that about 250 million people now living in those countries will eventually be killed by tobacco (Peto et al 1992).

Against this background, it may be that effects on health of a global environmental change would be largely undetectable for these diseases. (Infectious diseases are considered elsewhere.) At the very least, interpretation of epidemiological findings at a global (ecological) level is fraught with difficulty. Failure to appreciate these difficulties can result in what has been termed the 'ecological fallacy', such that inappropriate inference concerning aetiology at the level of individuals is made from the findings of ecological studies (Piantadosi et al 1988, Selvin 1958).

International collaborative epidemiological studies

A complementary approach to the international comparison of routine data is to collect and analyse data as part of a purpose-designed international collaborative study. Examples include the World Health Organization (WHO) MONICA (monitoring trends and determinants in cardiovascular disease) project for the registration of myocardial infarction (WHO MONICA 1987) and the EC-funded EUROCAT registry for congenital anomalies (Lechat et al 1985). Typically, such studies include a standardized protocol for the detection and registration of cases to reduce the problems associated with international comparison across routine registration systems. However, despite these efforts, problems may remain in some centres (Tunstall-Pedoe 1992).

The multinational design generally enables a wider range of exposure to the environmental agent of interest studied than is usually the case in a single centre, because countries with differing cultural, social and economic backgrounds are included: and, of course, the design can dramatically increase the sample size available for analysis. Furthermore, an internationally based project can study the effects of exposures that do not respect national boundaries. For example, the EUROCAT registry has been used to make an initial assessment of the effects across Europe of the Chernobyl accident (EUROCAT 1988).

The INTERSALT study. INTERSALT is probably the most extensive standardized international epidemiological study yet undertaken. It yielded unique and important data on the relations of sodium, potassium and other factors to blood pressure in 52 centres from 32 countries world-wide (INTERSALT 1988, 1989). Populations as diverse as the Yanomamo Indians of Brazil, Japanese insurance workers, Chicago businessmen and villagers in northern China were included. Up to 200 men and women aged 20–59 were recruited from random or whole population samples in each centre, giving an overall sample size in excess of 10 000 people. INTERSALT illustrates many of the advantages of carrying out international collaborative work for the study of the effects of environment (in this case, nutrition and other lifestyle factors) on health. In addition to generating a large overall sample size and hence high statistical power, the design of the study meant that it offered some important and unique advantages over the more conventional single-centre study (Elliott 1992).

(1) A wide range of sodium and potassium intakes was observed across the centres. Four remote population samples (the Yanomamo and Xingu indians of Brazil, Papua New Guinea highlanders and villagers from rural Kenya) had average 24-hour urinary sodium excretion (as a proxy for intake) markedly lower than those of the other 48 centres (INTERSALT 1988, Carvalho et al 1989). The highest sodium excretion and blood pressures amongst these four populations were recorded in the Kenyan sample (Fig. 2), which was in transition between its traditional rural existence and a more Westernized way of life.

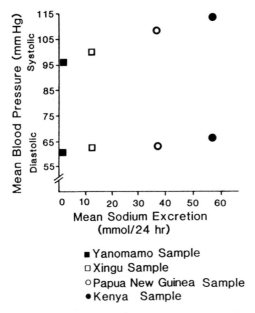

FIG. 2. Mean systolic and diastolic pressures against mean sodium excretion in the four remote population samples in the INTERSALT study that showed sodium excretion rates markedly lower than those of the other 48 centres. Reproduced from Carvalho et al 1989, with permission.

As soon as men from the area that provided the INTERSALT sample moved to Nairobi, their blood pressures were found to go up (Poulter et al 1990). These findings illustrate the relative importance of environmental over genetic factors in determining blood pressure distributions in populations.

(2) The international multi-centre design uniquely allowed the change of blood pressure with age, calculated from data on individuals *within* the centres, to be analysed in relation to average sodium excretion *across* the centres. Positive and highly significant associations were found both including and excluding the four remote low-sodium centres (which were found to strongly influence some across-centre analyses) (Fig. 3).

(3) The relation of sodium to blood pressure in individuals was analysed by multiple linear regression separately within each of the 52 centres, and then these 52 regression coefficients were pooled to give an overall estimate of effect. On average, a highly significant direct association was found between individual sodium excretion and systolic blood pressure (INTERSALT 1988), despite the well-known problem of 'regression dilution', which severely attenuates these within-centre associations because of large day-to-day variability in individual levels of sodium excretion (Elliott et al 1988). The distribution of regression coefficients relating sodium excretion to blood pressure *within* each of the

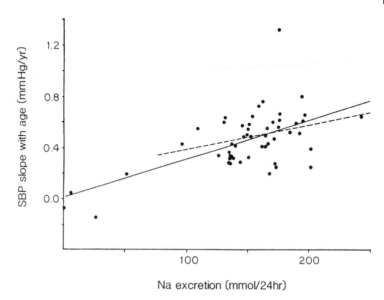

FIG. 3. Scatter plot of age–sex-standardized median systolic blood pressure slope with age (mmHg/year) against median 24-hour sodium excretion (mmol) and fitted regression lines. Solid line; regression line fit to the data from all 52 centres in the INTERSALT study (regression coefficient $b = 0.030$ [SE = 0.006] mmHg/year/10 mmol). Broken line; regression line fit to data set excluding the four low-sodium populations (see Fig. 2) ($b = 0.019$ [SE = 0.010] mmHg/year/10 mmol). Reproduced from INTERSALT 1988, with permission.

52 centres could also be examined *across* the centres (Fig. 4). This analysis indicated two centres that appeared to be 'outliers' in the tails of the distribution of regression coefficients. At the left end of the distribution a significant inverse association between sodium and systolic pressure was recorded for the centre in Osaka, Japan, which comprised insurance company employees and their wives. At the right end of the distribution a highly significant direct association was recorded in Nanning, People's Republic of China (Fig. 4). Findings such as these could suggest avenues for further enquiry, as discussed more fully elsewhere (Elliott 1992).

Studies of local effects: point sources of environmental pollution

Concern about the effects of environmental pollution on health among the public, the media and the scientific community alike has meant that apparent clusters of disease are more and more being brought to the attention of the public health authorities. Such clusters may be associated with point sources of pollution, often as an observation made *post hoc*. The problem facing public health physicians is how to investigate these clusters and how the results of any such investigation should be interpreted.

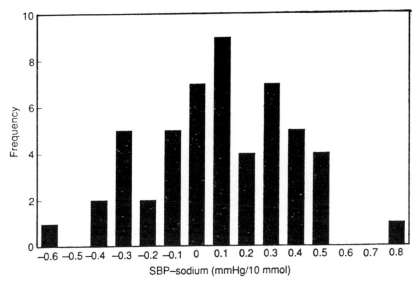

FIG. 4. Frequency distribution of the multiple-adjusted regression coefficients relating systolic pressure (mmHg) and 24-hour sodium excretion (per 10 mmol) in the 10 079 individuals in the 52 INTERSALT centres. The modal value is 0.1 mmHg/10 mmol sodium, uncorrected for the regression dilution bias. Reproduced from Elliott 1992, with permission.

Until recently, such investigations were costly and time-consuming because the data for any particular study had to be assembled in an *ad hoc* fashion. Often the area of concern crossed standard administrative boundaries, which complicated or made impossible the collection of the appropriate data. However, recent developments in computing, statistical methods and geographical referencing of health data have meant that the initial assessment of alleged clusters of disease near a point source can be completed rapidly, leading either to qualified reassurance (appropriate to the quality of available data) or to further study.

In the UK, the Small Area Health Statistics Unit (SAHSU) has been established with Government funding specifically to report rapidly on alleged clusters of disease, to study more generally the distribution of disease in small areas, and to advise on methods (Elliott et al 1992a,b). The SAHSU holds anonymized mortality and cancer registration data for Great Britain, as well as data on congenital malformations for England and Wales, and data on population at the small-area level provided by the national census. Geographical referencing of the health event data is via the unit postcode of residence: on average, a UK postcode relates to only 14 households. Map grid references exist for postcodes (accurate to 100 m) and directories are available to link postcodes (and hence the health data) to the underlying population data from census. SAHSU runs its own Unix-based super microcomputer.

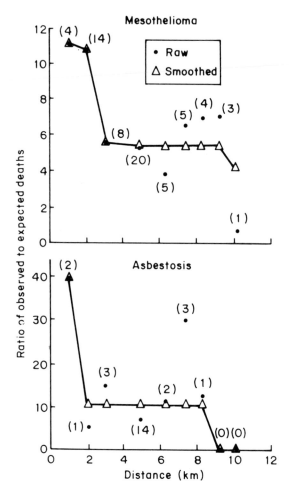

FIG. 5. Ratio of observed to expected deaths (O:E) in males 1981–1987 from *top* mesothelioma and *bottom* asbestosis against distance from Plymouth docks. The smooth line joins the estimated values of O:E subject to the constraint that risk declines with distance (according to Stone's [1988] model). Numbers of cases are shown in parentheses.

By using the postcode as a key, SAHSU can rapidly aggregate the relevant health and population data around any point in Britain. The observed numbers of events are then compared with the numbers expected on the basis of national or regional experience standardized for age and sex. Stratification by a measure of socio-economic deprivation of small areas based on census data is also carried out.

Figure 5 illustrates an example of the approach. Mortality from mesothelioma and asbestosis (an extremely rare cause of death) was studied for 1981–87 within a radius of 10 km of the Plymouth dockyards, where occupational exposure to asbestos was known to have occurred. Observed:expected ratios were calculated for a range of non-overlapping areas delimited by concentric circles up to 10 km around the docks. The data were analysed by a statistical model fit to the data which assumed that the risk either remained level or declined with distance from the docks, according to a modification of the method of Stone (1988). Observed:expected ratios for men are shown in Fig. 5, together with fitted values obtained from Stone's model. For men, highly significant excesses of both mesothelioma (based on a total of 64 cases within 10 km) and asbestosis (16 cases) were found overall, although there was no significant decline in risk with distance. No significant excess was found for women.

In the example illustrated here, the excess of disease in men was probably due to occupational rather than environmental exposure (Elliott et al 1992a). In another study, of cancer of the larynx and lung near incinerators of waste solvents and oils, no evidence of an association was found overall relating these diseases to proximity to the sites. The investigation followed reports of a cluster of cancer of the larynx around one of the incinerators (Elliott 1992c). Further studies underway include investigation of the effects on health of municipal incineration, vinyl chloride plants, benzene works and coke works.

Summary and conclusions

The study of the effects on health of environmental change is complicated by problems of data availability and quality. Routinely collected data on mortality and the incidence of cancers are of value for the study of international variations in disease risk and trends over time, provided caution is exercised in the interpretation of results. International collaborative epidemiological studies also offer a valuable method of studying the effects of the wider environment (including lifestyle factors) on health. Recent advances in methods for small-area studies have meant that the initial investigation of disease near point sources of pollution can largely be automated. More widespread use of these methods, to enable rapid replication of findings internationally as well as nationally, would enhance our ability to study the effects of environmental pollution on human health.

References

Anonymous 1992 The cancer epidemic: fact or misinterpretation? Editorial. Lancet 340:399–400
Barker DJP 1981 Geographical variations in disease in Britain. Br Med J 283:398–400
Breslow NE, Day NE 1980 Statistical methods in cancer research, vol 1: The analysis of case-control studies. International Agency for Research on Cancer, Lyon (IARC Sci Publ 32)
Breslow NE, Day NE 1987 Statistical methods in cancer research, vol 2: The design and analysis of cohort studies. International Agency for Research on Cancer, Lyon (IARC Sci Publ 82)

Carstensen B, Jensen OM 1986 Atlas of cancer incidence in Denmark 1970–79. Danish Cancer Registry, Danish Cancer Society and Environmental Protection Agency, Copenhagen

Carvalho JJM, Baruzzi RG, Howard PF et al 1989 Blood pressure in four remote populations in the INTERSALT study. Hypertension (Dallas) 14:238–246

Doll R 1991 Progress against cancer: an epidemiological assessment. Am J Epidemiol 134:675–688

Doll R, Peto R 1981 The causes of cancer. Quantitative estimates of avoidable risks of cancer in the United States today. Oxford University Press, New York

Elliott P 1992 Design and analysis of multicentre epidemiological studies: the INTERSALT study. In: Marmot M, Elliott P (eds) Coronary heart disease epidemiology: from aetiology to public health. Oxford University Press, Oxford, p 166–178

Elliott P, Forrest RD, Jackson CA, Yudkin JS 1988 Sodium and blood pressure: positive associations in a North London population with consideration of the methodological problems of within-population surveys. J Hum Hypertens 2:89–95

Elliott P, Westlake AJ, Hills M et al 1992a The Small Area Health Statistics Unit: a national facility for investigating health around point sources of environmental pollution in the United Kingdom. J Epidemiol Community Health 46:345–349

Elliott P, Kleinschmidt I, Westlake AJ 1992b Use of routine data in studies of point sources of environmental pollution. In: Elliott P, Cuzick J, English D, Stern R (eds) Geographical and environmental epidemiology: methods for small-area studies. Oxford University Press, Oxford, p 106–114

Elliott P, Hills M, Beresford J et al 1992c Incidence of cancers of the larynx and lung near incinerators of waste solvents and oils in Great Britain. Lancet 339:854–858

English D 1992 Geographical epidemiology and ecological studies. In: Elliott P, Cuzick J, English D, Stern R (eds) Geographical and environmental epidemiology: methods for small-area studies. Oxford University Press, Oxford, p 3–13

EUROCAT Working Group 1988 Preliminary evaluation of the impact of the Chernobyl radiological contamination on the frequency of central nervous system malformations in 18 regions of Europe. Paediatr Perinat Epidemiol 2:253–263

Gardner MJ, Winter PD, Taylor CP, Acheson ED 1983 Atlas of cancer mortality in England and Wales, 1968–1978. Wiley, Chichester

INTERSALT Cooperative Research Group 1988 INTERSALT: an international study of electrolyte excretion and blood pressure. Results for 24-hour urinary sodium and potassium excretion. Br Med J 297:319–328

INTERSALT Cooperative Research Group (Elliott P, ed) 1989 The INTERSALT study: further results. J Hum Hypertens 3:279–408

Kelson M, Farebrother M 1987 The effect of inaccuracies in death certification and coding practices in the European Economic Community (EEC) on international cancer mortality statistics. Int J Epidemiol 16:411–414

Kemp IW, Boyle P, Smans M, Muir CS 1985 Scottish cancer atlas 1975–1980. International Agency for Research on Cancer, Lyon (IARC Sci Publ 72)

Kleinjans JCS, van Maanen JMS, van Schooten FJ 1993 Human respiratory disease: environmental carcinogens and lung cancer risk. In: Environmental change and human health. Wiley, Chichester (Ciba Found Symp 175) p 171–181

Lechat MF, de Wals P, Weatherall JAC 1985 European Economic Communities concerted action on congenital anomalies: the EUROCAT project. In: Marois M (ed) Prevention of physical and mental congenital defects. Part B: Epidemiology, early detection and therapy, and environmental factors. Alan R Liss, New York, p 11–15

Lopez AD 1992 Mortality data. In: Elliott P, Cuzick J, English D, Stern R (eds) Geographical and environmental epidemiology: methods for small-area studies. Oxford University Press, Oxford, p 37–50

Mackenbach JP, Van Duyne WMJ, Kelson MC 1987 Certification and coding of two underlying causes of death in the Netherlands and other countries of the European Community. J Epidemiol Community Health 41:156–160

Muir C, Waterhouse J, Mack T, Powell J, Whelan S (eds) 1987 Cancer incidence in five continents, vol 5. International Agency for Research on Cancer, Lyon (IARC Sci Publ 88)

Percy C, Muir C 1989 The international comparability of cancer mortality data. Results of an international death certificate study. Am J Epidemiol 129:934–946

Peto R, Lopez AD, Boreham J, Thun M, Health C Jr 1992 Mortality from tobacco in developed countries: indirect estimation from national vital statistics. Lancet 339:1268–1278

Piantadosi S, Byar DP, Green SB 1988 The ecological fallacy. Am J Epidemiol 127:893–904

Poulter NR, Khaw KT, Hopwood BEC et al 1990 The Kenyan Luo migration study: observations on the initiation of a rise in blood pressure. Br Med J 300:967–972

Selvin HC 1958 Durkheim's 'suicide' and problems of empirical research. Am J Sociol 63:607–619

Stone RA 1988 Investigations of excess environmental risk around putative sources: statistical problems and a proposed test. Stat Med 7:649–650

Swerdlow AJ 1986 Cancer registration in England and Wales: some aspects relevant to interpretation of the data. J R Stat Soc A 149:146–160

Teppo L, Pukkala E, Hakama M, Hakilinen T, Herva A, Saxén E 1980 Way of life and cancer incidence in Finland. A municipality-based ecological analysis. Scand J Soc Med Suppl 19:1–84

Tunstall-Pedoe H 1992 Monitoring coronary heart disease in the community: why and how? In: Marmot M, Elliott P (eds) Coronary heart disease epidemiology: from aetiology to public health. Oxford University Press, Oxford, p 463–481

WHO MONICA Project Principal Investigators 1987 The World Health Organization MONICA project (monitoring trends and determinants in cardiovascular disease): a major international collaboration. J Clin Epidemiol 41:105–113

DISCUSSION

James: In the INTERSALT study, a complete outsider who knows nothing about these issues, would look at those data and think you have had to do an enormous amount of thrashing around to find a relationship. You have had to use huge data sets and enormous effort in terms of quality control, and have then come out with what appears to be a poor relationship. If we are dealing with general environmental factors, trying to look at a process where perhaps you can't even measure the events as well as you can blood pressure, then aren't we actually beginning to challenge the whole basis of environmental health analysis by the demands that you are making? I would guess that it's going to be even more difficult to show relationships than it was in the INTERSALT study.

Elliott: I agree. We are talking about the problems of measurement and the use of proxy measures of exposure. For example, in INTERSALT we were looking at sodium in a cross-sectional study. For most people we took a single 24-hour urine measurement, which we know is an extremely poor indicator of an individual's average sodium intake, and we used this as a proxy for life-long exposure to sodium. It is not surprising that the correlations we found appeared deceptively low. The problem is the same when we investigate the effects of physical agents in the environment, where again we're looking at poor measures, even no measures, and at best proxy measures of exposure. One may look at cellular effects, as we heard from Professor Kleinjans, which may be down the line of the true exposure. We are probably at the limits of epidemiology, and it's a challenge for us to get better measurements, more reliable measures and to validate them, and large studies are needed for this.

James: You are also running a risk, if it is a risk, of undertaking studies in which if you don't understand the variability and specificity of your measure you will get a reassuring negative result which is actually false.

Elliott: You need to understand and quantify what you are measuring. With biomarkers, for example, you need to know whether they vary, how reliable they are and whether they have any validity before using them as proxy measurements of exposure.

James: The trade-off between gaining specificity of measurements and the size of the sampling frame in terms of what one can afford to do is well known. At what level is the trade-off in environmental studies?

Elliott: There are essentially two approaches. Either you get a good and reliable measurement for the individual, or you get a lot of individuals. In INTERSALT we knew we had a poor measure (a single 24-hour urine collection from each individual), but we were at least able to estimate how reliable that measure was. We know that our initial results were highly biased towards the null value, but then, using data from repeat urine collections, we were able to quantify the extent of that bias and correct for it. So, we were able to make an informed judgement as to the correct nature of the sodium–blood pressure relationship and what it meant for public health. I think that this is a very instructive way forward. We need to understand what the problems of measurement are, estimate the variability, and then use these estimates to make judgements about the size of any effect. However, it is difficult in population samples to get lots of biological samples and get people to come back for repeat measurements. You are restricted in the sorts of things you can measure.

Klein: I can see one problem in your INTERSALT study and in the MONICA study, with which my institute is involved. The MONICA study is costing a few million DM per year and is running for 15 years. The effort is tremendous and the question is, what will the outcome be? Will we be able to influence the intake of the sodium when there are so many routes of entry, through eating and drinking, and so much geographical variability?

Why did you choose to look at sodium, when there is no way of altering its intake?

You spoke of environmental factors—an increase in blood pressure with age, for example—but you certainly didn't mean environmental pollution. The behavioural characteristics of people going into business and becoming stressed may also be relevant. The word 'environmental' should be explained and used carefully so that we do not mix up the things we have been discussing here, such as global climatic change and industrial pollution, with behavioural and other internal factors.

Elliott: The definition of environmental epidemiology that I and other environmental epidemiologists like is one that is broader than simply the study of the effects on health of chemical and physical agents in the environment, because one has to look at the wider picture. For example, we have data showing 2–3-fold differences in risk across Britain for lung and stomach cancer related to small-area measures of social deprivation. You have to look at the effects of environmental pollution over and above these socio-behavioural effects— they have to be part of your definition of environment.

INTERSALT is different from MONICA; MONICA is primarily a monitoring and registration study whereas INTERSALT was set up specifically as an analytical study designed to answer the question of whether blood pressure is related to sodium intake and other factors. I would say that it's been an important plank in the evidence on the relationship of sodium to blood pressure. It is not the only evidence—there is concordant evidence from animal studies, from clinical observations, from randomized controlled clinical trials, and from other within-population epidemiological studies. Even the sceptics now accept that sodium has something to do with blood pressure: there has actually been a large change of opinion over the last 5–6 years.

Now, what can we do about it? We can do a lot. Sodium is in our diet in such high amounts not because we choose it to be, but because it is put into food by manufacturers, particularly into bread and meat products. Around 75% to 80% of our intake of sodium is from manufactured food. A simple measure such as taking sodium out of bread, or reducing the amount of sodium in bread, will significantly reduce sodium intake. INTERSALT also looked at other factors which had statistically significant effects on blood pressure independent of sodium. We found an inverse association with potassium, a direct association with body weight and a direct association with heavy alcohol drinking (three to four drinks per day). With realizable changes in the combination of these factors in favourable directions, one can estimate a reduction in average systolic blood pressures of 5 mmHg or more, which, in public health terms, could have a huge effect on morbidity and premature mortality from coronary heart disease and stroke.

Edwards: Ten years ago I was involved in a genetics group at WHO. People wanted to do studies on sodium like the INTERSALT study but they were always opposed, on what I considered good grounds. The WHO was supposed to be worrying about the health of the world, and particularly of those who are impoverished rather than over-fed. Also, some of us considered that there was nothing which could be done on a global scale which couldn't be done in Boston, for example, to see if there was any major effect, and there were then serious doubts about this. The third problem was that once large studies have started they are difficult to stop, and tend to become more and more complicated and expensive. Obviously, the INTERSALT study has been beautifully analysed, and now it has taken place, I can't see how you could have got more out of it. But, if one is considering the practical point of the Third World, one has to ask, should it have started, and given that you have got so far, should it now stop? We have to allocate limited resources to world health problems, and other problems, including starvation and population control, are being virtually ignored.

Elliott: Blood pressure-related diseases are not only problems of the developed world. We had great trouble finding remote populations because they are disappearing. Once people are exposed to a 'Western' culture and lifestyle their blood pressure goes up very quickly and this is associated with much higher sodium in their diet and lower potassium and all the other things that go along with urbanization. When the Luo in rural Kenya went to Nairobi, their blood pressures went up very quickly, within weeks (Poulter et al 1990). This is not at all irrelevant to the developing world. It's fair to say that in developing countries coronary heart disease and stroke may initially be problems of the affluent, because the poor have to move to be exposed, but then all the evidence suggests that they become problems for rich and poor alike. On a global scale, the highest morbidity and mortality from these diseases will not be in the developed world.

Edwards: The association with sodium could still have been resolved in Boston, for example.

Elliott: Ten years ago people dismissed the evidence on sodium and blood pressure from within-population studies, studies of individuals. They said that one could find these effects across populations but not within populations, because of all the problems that Philip James raised of measurement and proxy measures. You need a big sample and very good methodology to study such effects. I don't think it's true to say that it would have been easier to do a comparable study in Boston where you would have needed to recruit 10 000 people. There are problems of standardization and quality control. We devolved the data collection to 52 centres, each of which had to get data on 200 people. If we had asked them to study 400 or 500 people it would have worked less well; 200 seemed to be about right. That study design is now being used successfully in at least three other major international studies using the multi-centre approach. Also, we had the advantage of being able to look at effects across the centres; for example, the findings on change in blood pressure with age are unique to INTERSALT and extremely important. Within one study, we were able to look for generalizability of effects in many different places. If you carry out your study in one centre, looking at Whitehall civil servants for example, people ask how your findings relate to people in Brazil or China or Japan. INTERSALT is a global study and our best estimates suggest these results are generalizable, relevant not just to Boston but also to the Japanese, the Chinese, the Kenyans and to Europeans. Finally, on a practical note, the central costs of the study were relatively low, because each of the centres raised its own money; they had to provide the staff and the finances to enable them to collect data from 200 people and to send the data to be analysed centrally. The true, total, cost of the study was large, but it was a manageable cost centrally and for each of the participating centres.

Edwards: The true cost is the total cost involved. I don't doubt that the study is an absolute model of how to do things, but it does seem to be a strange target on which to practice your marksmanship.

Elliott: I disagree. In cardiovascular disease, next to smoking, diet is the most important environmental exposure (under my broader definition of the environment). It's extremely difficult to study, but that doesn't mean that it shouldn't be studied. It can be studied well and properly, but standardization, quality control and large sample sizes are needed. A multinational collaborative study is an attractive and effective approach.

Bobrow: Epidemiology is one of many ways of studying biomedical problems and it's good for some things and not good for others. There is an important dichotomy when considering environmental monitoring and regulation as to whether the triggers should be based primarily on observed health effects, or on physical measurements in the water, soil, etc. Obviously, the two interdigitate, but I think you can separate them. Within this meeting, some people say regulations should prevent the level of contaminants reaching a point at which there is an observable effect on human health. Others say that the regulations should prohibit any contaminant in drinking water. This leads to the fundamental problem of how to decide what should not be there. This is not always easy. I believe what is ultimately important is human health. We must be prepared to invest in the development of better methodologies for measuring human health, and for getting around the current disadvantages of using health measures as a monitoring mechanism—the high cost, insensitivity and long latency periods. If, on the other hand, you believe that all 'contaminants' must be eliminated, regardless of whether or not they have effects on health, there's no point investing in setting up complex epidemiological monitoring systems.

Reference

Poulter NR, Khaw KT, Hopwood BEC et al 1990 The Kenyan Luo migrant study: observations on the initiation of a rise in blood pressure. Br Med J 300:967–972

Environmental and health problems of developing countries

David J. Bradley

Department of Epidemiology & Population Sciences and Ross Institute, London School of Hygiene and Tropical Medicine, Keppel Street, Gower Street, London WC1E 7HT, UK

Abstract. Environmental variables largely determine the disease pattern in developing countries. Infections and malnutrition predominate, due to the effects of both poverty (a summary of many aspects of material deprivation) and a high ambient temperature. Environmental changes may be intended to improve health—examples include improved domestic water supplies and sanitation—or they may be due to socio-economic developments, which often have favourable or unfavourable health consequences. These are explored for water resource developments, where the health effects are complex, and for deforestation and urbanization. Although environmental impact assessment has been of value in reducing the adverse health impacts of socio-economic development projects, the use of health opportunity assessment is proposed as a more positive approach to optimizing the consequences of development to human health.

1993 Environmental change and human health. Wiley, Chichester (Ciba Foundation Symposium 175) p 234–246

The rate, diversity and complexity of environmental change today is particularly great in developing countries. This paper reviewing some of these changes complements that earlier in this volume (Bradley 1993). The first paper addresses the definition of environment, the relation between developing countries and a tropical location, and the health consequences of change. It then deals in more detail with the direct effects of global climatic change upon the transmission of vector-borne diseases, with special reference to malaria. It has a specific disease focus. This paper goes on from the general aspects of the first paper to look, with an environmental focus, at the changing pattern of health and disease.*

The environment, as broadly defined (Bradley 1993, p 148), has a massive impact on health and disease in developing countries, and until recently has

---

*This paper replaces that which was to have been given by Professor Alvaro Umaña, who was unfortunately unable to attend the symposium at the last minute.

dictated their demographic structure also. At this symposium, the papers about the European environment have sometimes appeared to reflect the 'delusion of arcadia'; that is to say that the European approach to environment treats the natural world in terms of the addition of pollutants to it. In the Third World we have to deal with a very heavy environmental determinism, starting from a situation in which things are very far from what they should be. A second broad generalization is that aspects of the environment concerned with the limitation of resources may be of greater importance to developing countries than the pollution of the environment. In developing countries pollution, from the point of view of human health at any rate, is mainly organic and microbial rather than chemical, except where urbanization and economic growth are substantial. A fourth aspect, that of population growth neutralizing progress, was dealt with thoroughly by Dennis Lincoln elsewhere in this volume (Lincoln 1993).

There has been much discussion in international health circles about the demographic transition, the idea of moving from a high birth rate, high death rate society to a low birth rate, low death rate society. There has more recently been discussion of the concept of an epidemiological transition, in which there have been even more gross oversimplifications, involving a fall in the diseases caused by infection and parasites and malnutrition, and their replacement by what some people call diseases of lifestyle, cardiovascular disease and cancer, although these could be regarded as being just as much environmental as the former group. The substantive point in both the demographic and the epidemiological transition concerns the relative timing of the two components of each. In demography, the gap between the fall in death rates and that in birth rates will determine the amount and rate of population growth. In the so-called epidemiological transition, will the infections decrease before or after the rise in 'diseases of lifestyle'? In many situations, especially among the peri-urban poor, the latter group of diseases (and cancer is common in some industrially undeveloped societies) rise much earlier than the former group come down; there will therefore be a period when although things may have improved in terms of death rates, the environment and the community are faced with the worst of both worlds. However, in a poor rural developing country the disease pattern is dominated by infections and malnutrition. In Guatemala, for example, a typical village child scarcely increases in weight between the ages of six months and three years, because he is subject to a vast number of attacks of diarrhoeal disease and a great variety of other infections. The starting point is not one of an unpolluted world, but one in which malnutrition and infections dominate the scene. These problems fall into essential two groups. There are the warm climate diseases discussed earlier (Bradley 1993)—those involving stages outside the human body which are temperature dependent and so climatically dependent and environmentally determined—and those due to the complex of things which one can lump together as poverty. In many Third World countries the people seen at a clinic will resemble those seen in a poor part of Europe a hundred years ago.

As one moves from poorer to richer parts of the world, one sees a decline in the importance of infections and an increase in the importance of cardiovascular disease. However, injuries loom very large in both rich and poor societies, although the nature of the injuries varies from country to country. A dichotomous view of the world is, however, really much too simple.

Health consequences of environmental changes within developing countries

Among the many environmental changes relevant to socio-economic development, I will emphasize water resource development and urbanization (Weil et al 1990), looking first at the traditional concerns of environmental health in the narrow sense: domestic water supply and sanitation.

Domestic water supplies

It has been argued that over 50% of illness in developing countries is water-related. Domestic water supplies are particularly appalling in arid areas and urban slums, with deficiencies involving both quality and quantity. Where water has to be carried from a source outside the compound, per capita use rarely exceeds 20 l/person/day, which is below the 30 l believed necessary for reasonably hygienic behaviour (White et al 1972). In spite of substantial efforts to improve the situation, poor maintenance of domestic supplies and population growth have meant that the number of people with seriously defective supplies remains large. Organic pollution with human wastes remains the chief health hazard from water quality defects.

There is enormous variation in the amount of water families use, even within East Africa. A multicentre study within three East African different populations, both urban and rural, revealed that people in towns use much more water than people in the country, and that those who have to carry water use much less than those who have it piped to their houses. This raises the question of how much water one needs for health as distinct from other purposes. I refer to domestic water, not water for agriculture or farming or for stock. There has been a shift over the last two or three decades from the European preoccupation with quality of water to a consideration of the availability of water and the relationship between the two.

In urban water supply, the key finding that emerges from any field study is the high cost of water to people with a low income. In Uganda it is common for unskilled urban workers to be spending 10% of their income on acquiring very poor quality water in quite limited quantities. In contrast, someone in the USA will need about 1–2 days' income a year to provide an abundant water supply. There is a large list of diseases associated with water. If one presents this to engineers, they switch off very rapidly, understandably, and if one starts telling them all about the biology they lose interest even faster! A great deal of progress was made in communication, and also in the way we conceptualize the problems of domestic water, when the problems were categorized not in terms of the biology of the agent, which is the traditional way, but in terms of the intervention that would change the situation. This revealed (Bradley 1993, White et al 1972) that there are four groups of infectious diseases related to water—the diseases in which the water carries the infectious agent, those resulting from the absence of a water supply sufficient for adequate personal hygiene and personal cleanliness, those diseases transmitted by intermediate hosts such as snails which live in the water, and the vector-borne diseases carried by insects which are related to water. These groups require quite different approaches to prevention, and it is also important to look at their relative importance. One thing that emerged from a whole series of studies in the late 1960s and early 1970s was the importance of water availability relative to water quality. In developing countries, it became clear that many of the diseases which people tend to think of as being water-borne were not actually water-borne, but were food-borne or directly faecally–orally transmitted, and what the people needed was more water for washing, even if it was somewhat polluted water rather than very high quality water. This supposed need for high quality water had in fact been holding up rural water supplies. After this change in perception of the issue, a substantial amount of money, of the order of several hundred million dollars at the very least, was moved by the World Bank into rural water supply.

In the 1970s there was a tendency on the part of public health engineers to view the world in dichotomous terms: that either you could do nothing about water supply in a developing country, or people had to have a properly filtered and chlorinated multiple tap supply. It soon became clear that there was a whole variety of intermediate measures varying from untreated roof catchments, through municipal stand-pipes to various forms of rural pipe-lines which might supply water of variable quality. In particular environments, by breaking the situation down and looking more flexibly at the options and at the costs of particular interventions and the benefits in health terms, to some degree on a best guess basis, it became possible to find low cost alternatives which would reduce disease substantially. One of the major conclusions, which seems to have been generally accepted, was that it is crucial to bring water to the compound. If the water is available on the compound, even in small amounts with a flow-limiting device, usage will go up to about 30 l/head/day or more, and most of the health benefits of domestic water can probably be gained at that level of use. It is not necessary to go up to the enormously high levels which are customarily used when water is readily available at multiple outlets within the home.

To prevent some other types of water-related disease it is necessary to find ways of preventing people from going into the source, and various other *ad hoc* approaches may be used. The key point is that policy options, the approaches, need more basic thought and a greater range of options than we tend to think of in terms of our traditional guidelines.

Water resource developments

Separate from domestic water supply improvements are water resource developments for other purposes, and another group of issues affect these. When one builds a dam, a whole series of things happen. In Europe it is possible to think of a particular heavy metal pollutant in isolation from others or with just another one or two, whereas most developments in Third World countries take place as developments of a whole habitat or lifestyle; creating an irrigation scheme or a large dam involves moving people, changing the availability of water and changing the nature of the soil that is being cultivated. Although it is useful analytically to think of the elements, in terms of policy one has to look at the 'package'. Because many of these packages are in fact created in the form of a specific project which involves the lending of money, there is one time, and I think only one, at which you can *really* influence the environmental issues. This opportunity lies with the person who writes the terms of reference for the engineers or whoever will be dealing with the socio-economic development. At this point, one can influence the priority that will be attached to health and of environmental issues, and if they're not included at that stage, there is almost no chance of them being put in later on. Large water resource developments also produce extremely complex environmental changes.

Water impoundments

In the tropics massive reservoirs and a vast number of small impoundments have been built to provide hydroelectricity, surface water supplies for towns and stored water for irrigation, and for flood control. A large literature exists on the big impoundments (Bradley 1977). These were typically constructed, one per country, in the early post-independence era of formerly colonial tropical countries. As an example, Volta Lake in Ghana is vast, with an irregular periphery providing over a thousand miles of shoreline. Construction of such reservoirs leads to changes in vector ecology and also involves extensive human migration. The former inhabitants of the lake bed are displaced and, usually unsatisfactorily, resettled, and construction workers move in and later depart. Planned immigrants usually arrive to farm the irrigated lake periphery, and unplanned immigrant fisherman arrive in large numbers soon after inundation to exploit the often eutrophic waters. The breeding places of *Simulium*, vectors of onchocerciasis (river blindness), are flooded out, but the spillway of the new dam may be recolonized by *Simulium*. The extensive still water, especially if the sides slope gently, provides an ideal habitat for *Bulinus* and *Biomphalaria*, intermediate host snails for schistosomiasis, and epidemics of this disease follow; it may then become endemic and prevalence of the disease may reach very high levels. Transmission may be kept down by carefully planned engineering of the banks and shoreline. Long-term secular changes may be involved. In Lake Volta these extended over 15 years. Initial inundation was followed by death of trees in the flooded area, massive proliferation of wood-eating insects, huge increases in the fish population and massive immigration of fishermen to exploit a catch that rose to a peak of 60 000 tonnes per year. The fishermen's extensive contact with water coincided with huge increases in *Bulinus truncatus rolfsi* on the water weed that had appeared because of the eutrophication of the water resulting from tree rotting, and extensive outbreaks of schistosomiasis haematobium occurred. The epidemiology was complicated by annual migrations of the fishermen back to their villages of origin (where there was also some local transmission of schistosomiasis) several hundred kilometres away for a month each year for festivals. Once the trees had completely rotted away, insect populations fell, the fishery decreased and some fishermen departed, there was less eutrophication, less water weed and a reduction of schistosome transmission.

The precise consequences of impoundment will depend on the ecology of the local vectors, as Lake Volta again illustrates. It had originally been suggested that a huge siphon should be built into the Volta dam so that the water level could be abruptly dropped once a week, the aim being to leave larvae of the anopheline mosquito vectors of malaria stranded. This idea was suggested by the highly successful use of siphons in the impoundments of the Tennessee Valley Authority. The concept proved impossible to apply on the scale of the Volta dam, which was fortunate, because the chief African malaria vector, *Anopheles gambiae*, is a puddle breeder with a short development time, and the proposed variation in water level would have greatly increased the population of this mosquito. Even this would not have had the effects that might have been predicted, because baseline transmission before the dam was built was already so high that increasing anopheline populations would have had little effect on the amount of malaria.

The complexity of the epidemiological consequences of creating this single large reservoir, in a relatively well-studied area, should warn against simplistic predictions of the health consequences of any environmental change.

In the rapids at the source of the Nile from Lake Nyanza in Uganda the main vector-borne disease problem was that of river blindness carried by *Simulium* breeding in the aerated waters of the rapids. When the dam was built, the rushing waters of the rapids were exchanged for still water, and the problem switched to one of schistosomiasis, which is easier to treat. Also, the chief vector control officer of the Ugandan government insisted that a small pipe be put into the dam wall when it was built, down which 500 gallons of DDT (dichlorodiphenyltrichloroethane) could be poured. With the turbulence of the water and the right choice of dispersants it is possible to produce 1 μm particles of DDT, which are selectively filtered out by the *Simulium*. This cleared *Simulium* for 70 miles downstream without causing major changes to the rest of the biota, reducing river blindness enormously and making the area habitable. In fact, when this was stopped because of civil strife, we suspect that another strain of *Simulium*, which does not feed on humans, invaded, so the problem may have been solved permanently.

Construction of large dams usually requires international funding, which provides the opportunity for an organized process for ensuring that environmental and health problems are minimized. The difficulty is that in most countries, many more people are influenced by small scale changes. In Nigeria there is one big dam at Kainji, whereas in even one small area of northern Nigeria more than 500 small dams have been created. The issue of how to plan and design measures to deal with health problems and to prevent them arising under these circumstances has not been satisfactorily addressed.

Irrigation

Irrigated agriculture has been greatly expanded in developing countries to bring new areas into production and to move from situations in which there is annual uncontrolled flooding to one in which water supplies are perennially available and controlled. The effects on vector-borne diseases resemble those of impoundments, but man–water contact becomes much closer and most of the water flows slowly. Schistosomiasis is the main problem because the canals, and even more the drains, provide good habitats for snails, and both water pollution by excreta and water contact with the skin are hard to avoid. Malaria may also increase substantially, especially in areas where transmission was previously low or the disease was epidemic. However, local variation is usual. In The Gambia, for example, irrigation of land to grow rice led to a second peak of *An. gambiae* in the dry season, but this was not accompanied by a second rise in malaria cases, for reasons not yet fully understood.

Some of the most extensive areas of (temporary) fresh water in the world are irrigated rice fields. These may act as breeding sites for many vectors, occasionally of malaria (*An. sinensis* in China) and more often those of Japanese encephalitis B (*Culex tritaeniorhyncus*), a potentially lethal infection that destroys mental function in many of those it does not kill. Detailed examination of breeding of anophelines in rice fields often shows that much of the breeding takes place in the culverts and areas concerned with the irrigation system rather than in the rice fields themselves. By pin-pointing where the problem lies, one can deal with it by environmental means relatively successfully.

Timing and phenology are also important. In some areas rice fields are important and increasing in numbers. The snails that transmit schistosomiasis live in rice fields. However, examination of the agricultural cycle shows that in Tanzania by the time the extrinsic cycle of the parasite has been completed in the snail the rice field has dried up, so there can be no transmission. It is important to look at all aspects of transmission and not simply aim to kill all snails in rice fields.

Where new strains of high-yielding rice are introduced, the use of fertilizers and insecticides needs to rise, and ecological diversity in the rice fields is reduced. Where previously many edible crustaceans and non-rice plants developed and were eaten freely by the population, there now is a productive monoculture. Aggregate wealth may rise, but the landless poor are now deprived of several food sources and may be even worse off.

Land development: deforestation

The removal of forests affects malaria and other vector-borne diseases both in Latin America and in southern Asia. In Brazil, those constructing roads through rain forest have suffered from mucocutaneous leishmaniasis transmitted as a zoonosis from arboreal and other small mammals to the intruding people by forest sand-flies. Malaria is a hazard not only to these road-makers but also to agricultural populations pushing back the forest edge. The most effective vector, *An. darlingi*, is a mosquito of the forest and forest edge and the incidence of malaria is very high, especially where the land is relative low-lying.

In southern Asia the problem of forest-fringe malaria extends from India to the Far East. The problems have been well studied in Thailand where transmission is by *An. dirus* (formerly known as *balabacensis*), which is a forest species, and by *An. minimus* in the cleared areas. The forests cover the hilly areas near the national boundaries. Illegal tin-mining and gem-mining create multiple small water-filled depressions suitable for vector breeding. The illegal activities, including logging and smuggling, make for lawlessness and limit the government's malaria control activities, and these areas tend to be inhabited by ethnic minorities who do not interact comfortably with the government. This pattern, with local variations, extends along much of the border areas of South-east Asia. Most falciparum malaria in Bangladesh, for example, is from the forest fringe areas of the Chittagong Hill Tracts.

One aspect of development and of environmental contamination needing emphasis, but not addressed elsewhere in this volume, is that of changing the fauna and flora, introducing new plants and animals without adequate attention. Sleeping sickness, transmitted by tsetse flies, killed a third of the population of Uganda at the beginning of this century. *Lantana camera* was introduced into western Kenya by an ill-informed agricultural officer who thought it would make a nicer fence round the fields than did the hedges of *Euphorbia*. This enabled the tsetse flies to leave the lakes and proliferate near houses, resting in the highly humid *Lantana camera* bushes. Six hundred cases of sleeping sickness followed in a few years in one small district.

Urbanization

The rate at which populations in developing countries are becoming urban is great, with major cities often showing growth rates of as much as 9% per year (as in Bogota) as a combination of immigration and reproduction. In general, urbanization and especially the lives of the urban poor are marked by problems of domestic water supply and the removal of waste water and excreta, and rates of diarrhoea are high. Vector-borne diseases may be lower than in the rural areas, but are not always. Malaria in Indian cities, transmitted by *An. stephensi*, a highly effective urban vector, may be severe.

Cities generate more polluted and less clean surface water. This is associated with a swing from anopheline to culicine mosquitoes, especially *Culex quinquefasciatus*, which breeds in flooded pit latrines and in polluted ditches where the density of ovipositing females may be high enough to totally cover the water surface! They act as vectors of Bancroftian filariasis (elephantiasis), which has moved from being a rural to an urban health problem.

The majority of urban immigrants arrive poor and without employment, and consequently settle as squatters on marginal land lacking urban services. These densely settled areas are often on steep hillsides prone to floods. The interaction of all these adverse environmental factors with social and economic deprivation and overcrowding lead to poor health from a combination of the diseases of poverty with the classical tropical diseases.

Reducing environmental and health problems

In many situations in developing countries, particularly in internationally financed projects, concerns about health have arisen as a side effect of environmental concerns. Some years ago, the World Bank was not concerned about health, except when disease was an environmental consequence of its projects. The World Bank's emphasis in health was on schistosomiasis for a time, because they were building irrigation schemes which led to an increase in snails and therefore an increase of schistosomiasis. The health issues arose as a consequence of that sequence. This was disadvantageous from the point of view of health, if not from the point of view of environment. In the Philippines, the rehabilitation of a national irrigation scheme was going to involve expenditure of 120 US dollars per head to prevent the schistosomiasis situation being made worse by environmental changes. Nothing was being done about any other aspect of health. Given such a vast amount of money to spend on health, which would not be available again, there were far more efficient ways of improving health than worrying about schistosomiasis. In development projects, not only should the environmental impact be assessed, but also health opportunities provided by the project should be seized as an end in themselves, and health treated as more than subsidiary to the environment. The environment should be considered in its own right, but so too should health, as well as the interactions between them.

The broad issues include the following. One needs to distinguish between environmental change brought about in order to improve health, and environmental change aimed primarily at development, which can have secondary effects on health which may be highly beneficial or detrimental. There may be particular effects on diseases, or they may occur through economic changes. I am not happy with the view that one should look at the physical environment separately from the social and economic environment. A lot of damage can be done in developing countries if one looks at only one segment at once, because of the close interactions between them. In the same way, in developed countries behaviour and environment should not be looked at separately; here I find myself in agreement with only some of the views expressed at this meeting. Several other philosophical issues are relevant. The question has been raised as to whether we are looking primarily at health or primarily at a measure of something in the environment (p 233). If we are using an environmental guideline, which we have to do because we cannot measure health every time, are we using a 'standard' as a goal at which to aim, or something which must be attained. Should the World Health Organization's water standards be lowered in many countries in order to put pressure on people to attain them, or should they be raised to provide an ideal target?

How far should approaches be transferred from industrial countries? The treatment of sewage in developed countries is concerned with reducing biological oxygen demand and not with reducing pathogens. Pathogens are taken out of the water supply system through the treatment of water. In developing countries it is not always feasible to treat water supply systems. It is, however, much easier to reduce pathogens by means of oxidation ponds in many parts of the tropics than it is in temperate countries. The philosophy of where in the water cycle you remove pathogens may well be environmentally determined.

A final question is how far we are pessimistic or optimistic. There is a monastery in Thailand where the abbot has installed a biogas plant. People go round the surrounding village collecting up animal excreta and, to some extent, human excreta to put into it. The biogas plant produces methane, which is burnt and used to heat the rice to feed the monks and other people to whom they give the rice. This seems to sum up many of the goals of environmental health interventions in the Third World. It is an integrated thinking, both using a resource which would otherwise be wasted, and reducing pollution, organic pollution in this case. It is linked to social perception; something which was perceived as a waste product is now viewed as a resource because it can be utilized, when in most cultures it is very difficult to get people to worry about waste products. Also, this is a small scale development, reproducible with minimal expense on a large scale. For many parts of the Third World, particularly in waste disposal, both urban and rural, one can find a socially acceptable, relatively low technology method which can be replicated on a large scale indigenously and does not need to be driven by external forces from outside the country. This is perhaps the most hopeful approach. The contribution the developing countries can make to thinking in developed countries is in showing the need to go back to basics and not just to build on relatively simple rules of thumb as we traditionally have done.

The rate of induced environmental change in developing countries is extremely rapid. Some of this change is specifically intended to improve human health; most of it is primarily aimed at economic development, and may have some adverse consequences for health. These have sometimes been detected in advance by environmental impact assessments. A more flexible and effective approach will be to make a Health Opportunity Assessment of each development project at the planning stage.

Acknowledgement

I am grateful to Professor R. M. May and the Oxford University Zoology Department for the sabbatical environment in which this article was written. It forms part of the ODA Programme in Tropical Diseases Control.

References

Bradley DJ 1977 The health implications of irrigation schemes and man-made lakes in tropical environments. In: Feachem R, McGarry M, Mara D (eds) Water, wastes and health in hot climates. Wiley, Chichester, p 18–29
Bradley DJ 1993 Human tropical diseases in a changing environment. In: Environmental change and human health. Wiley, Chichester (Ciba Found Symp 175) p 146–170
Lincoln DW 1993 Reproductive health, population growth, economic development and environmental change. In: Environmental change and human health. Wiley, Chichester (Ciba Found Symp 175) p 197–214
Weil DEC, Alicbusan AP, Wilson JF, Reich MR, Bradley DJ 1990 The impact of development policies on health. WHO, Geneva
White GF, Bradley DJ, White AU 1972 Drawers of water. University of Chicago Press, Chicago, IL

DISCUSSION

Lake: You consider that urbanization will be of greater significance than total population growth. Dr Elliott, you described an effect of urbanization on systolic blood pressure. After you have allowed for the effects of salt, the negative correlation with potassium, and the positive correlation with body weight and with alcohol consumption, is there a residual urbanization effect? This might be difficult to detect in a large epidemiological study, but is there evidence from smaller studies for urbanization having an effect independent of increased density of population?

Elliott: We don't know all the factors that are involved in the rise in blood pressure observed when people move, though it's certainly a well-documented phenomenon. When whole populations move, as, for example, when Tokelamans moved to New Zealand, their blood pressures go up as a group (Prior & Stanhope 1980). The Kenyan study is the only one which has actually followed people systematically over time from their rural environment to an urban environment (Poulter et al 1990). Some but not all of the change in blood pressure could be explained in terms of things that could be measured, such as sodium or potassium intake. So, there is some unidentified effect of urbanization, or a change in circumstances, but we don't know what it is.

Edwards: Does malnutrition actually lower blood pressure?

Elliott: Chronic malnutrition can be associated with either high or low blood pressure, but in general wasting is associated with lower blood pressure.

Lake: Dr Bradley's concern seems to be that disease would be more rife in an urban situation. Intuitively, I would prefer people to be in rural communities, but we have to face the fact that that might not actually be the best situation.

Bradley: My concern is that the rate of growth of the cities may be too high. I wasn't trying to imply that urbanization worries me more than population growth; people move to cities because of population growth. It seems to me that a large city is a relatively unstable situation in terms of sources of food and so on, and is vulnerable to disruption. The big catastrophes resulting from population growth will probably happen in one or two of the large cities. In Mexico City, for example, the nearest water supply now is many hundreds of kilometres away—all the nearer supplies have been used up—and if more water is needed for more people, it will have to be brought from a very long way away. The complexity of cities makes them rather unstable for the future.

Bobrow: If you are right in saying that a poor urban immigrant will be worse off than he was back in the rural environment, what is it that drives people to continue moving into urban areas?

Bradley: Let me be clear. What I said was that it was believed previously that it was much more healthy to be absolutely poverty stricken in a city than in a rural area. The evidence now suggests this is not the case. The urban poor may be comparably badly off. Whether there is a gain for *most* people in moving to the city is not absolutely clear, but we are only just beginning to subdivide urban populations for demographic and health studies in sufficient detail. Previously, if you'd said that urbanization was heading for trouble, people would say, 'but look how much more healthy people are when they go to a city'; it isn't as clear cut as that.

Bobrow: Do you think this is because we are now learning what the truth has always been, or do you think there has been a change in the situation over the last five or ten years?

Bradley: I think it's probably what the truth has always been for a large city to which a lot of landless people are migrating. Secondly, there are more large cities now, and there are more people, so there are bigger flows.

Zehnder: There might therefore be an optimum size for a city. We should investigate what might happen if that optimum is exceeded, and what factors actually define the optimum size.

References

Poulter NR, Khaw KT, Hopwood BEC et al 1990 The Kenyan Luo migration study: observations on the initiation of a rise in blood pressure. Br Med J 300:967–972

Prior IAM, Stanhope JM 1980 Blood pressure patterns, salt use and migration in the Pacific. In: Kestelook H, Joossens JV (eds) Epidemiology of arterial pressure. Martinus Nijhoff, The Hague, p 243–262

Evolution and environmental change

John H. Edwards

Genetics Laboratory, Department of Biochemistry, University of Oxford, South Parks Road, Oxford OX1 3QU, UK

Abstract. The effects of agriculture have been to improve the stock by selective breeding at a necessary loss of some capacities to withstand epidemics, while movements of products and stock have increased the speed with which epidemics spread. Some implications are considered. In relation to human genetics there is also evidence of a lack of selection, rather than any directed selection, and this will necessarily have some long-term effects. However, they are not well understood and are certainly slight, but cumulative. All that can be done at present is to maintain the lowest practical exposure to mutagens: the best understood, radioactivity, does not appear to be a serious hazard in the absence of catastrophic events or inappropriate medical usage.

1993 Environmental change and human health. Wiley, Chichester (Ciba Foundation Symposium 175) p 247–259

Evolution is a change in composition of the genetic heritage of a species. In the natural state it is the consequence of responding to environmental change, partly due to the environment actually changing, but mainly to organisms colonizing new environments. Evolution is driven by the different reproductive rates of individuals with different genetic compositions, and may be regarded as having a cost in terms of the loss of potential mature individuals. For a given degree of evolutionary change, this cost will be inversely proportional to the difference between those genotypes which contribute and those which fail to contribute to the next generation. It is at best a slow procedure in the wild. The 'cost' of exchanging merely one gene for another may exceed the standing population several fold. The fossil record suggests changes in size and shape rarely exceed 1% per millenium in the wild. However, because in most organisms the chance of any gamete conveying an allotted half of its genes to two or more offspring is less than a half, substantial costs are being incurred anyway, so that very rapid changes can be achieved without any appreciable additional cost.

It might be assumed that if the breeding and non-breeding subsets of the population were randomly chosen evolution would stop in the absence of selection. This would be true if the genotype were stable, but it is not. Mutations are occurring constantly and, because the vast majority of these will be detrimental, selection in the form of an inequality of contribution to the next generation is necessary. If the proportion of new variants increased from generation to generation, because most are detrimental, there would be a steady degradation in those features which distinguish a species from its remote ancestors. This has been termed the 'Red Queen effect'; in the words of Lewis Carroll, it 'takes all the running you can do, to stay in the same place'. As an analogy, consider a manuscript successive editions of which are typed by illiterate typists with very occasional errors. The degradation of meaning will be slow and progressive, with a few errors providing new insights, or even new and useful words, to an informed reader.

I wish to discuss two separate issues of evolutionary change and stagnation: firstly, modification of the genotype by the selective breeding of those species on which we depend; secondly, the consequences of the various forms of environmental change likely to damage the genetic endowment of our own species and the feasibility, or rather unfeasibility, of repairing this damage to our genome by any substantial or acceptable laundering.

The consequences of selective breeding

There are over a million species but only a few dozen are domesticated and selectively bred to provide a substantial proportion of edible or wearable products. Over 90% of the world's food is derived from fewer than ten species, dominated by wheat, rice and maize and the secondary source—the cow, sheep, pig, goat and chicken—which convert leaves and seeds with an efficiency of between 10% and 20% and provide the extravagant luxury of varied and concentrated foods.

Wild species have evolved largely to avoid being eaten, from the outside by larger predators, by smaller predators such as blood suckers and surface parasites, and from the inside by bacteria and viruses. Higher plants cannot move, and, because they need light, cannot hide, and must rely on tough exteriors, thorns and a wide variety of toxins, varying from simple compounds such as cyanide to hallucinogens and proteins which specifically interact with the immunological system; as long-term aids, they use teratogens, carcinogens and mutagens. Animals, which eat plants or eat animals which do so, have a wide repertoire of strategies for survival; apart from the obvious behavioural ones of stealth, flight, cleanliness and selective diets excluding toxic plants, they have more subtle defences intended to ensure survival of some of any population. In humans, these include the ancestral immunological memory provided by some closely linked and functionally related loci in the main defence systems on which we are dependent for the exclusion or inactivation of bacteria and viruses. However, these recipes for tribal survival, by which some individuals will survive the worst epidemics, were mainly honed in the Stone Age and earlier, and can lead to problems in the present. As with any defence system, if it is not exercised and disciplined, apathy, mutinies and other unwelcome and inappropriate responses can arise. It is hardly surprising that the cost of eliminating and excluding various parasites imposes other hazards of immunological mutiny such as diabetes, rheumatoid arthritis and multiple sclerosis. Farm animals presumably suffer similar problems, and, if they do not, may well have lost some ability to respond to the unexpected.

Domesticated species are subjected to the obverse of artificial selection, and are encouraged to breed only if they or their close relatives appear edible, or produce such useful products as milk or wool of above average value.

In a few thousand years of informal selection these species have distanced themselves from their ancestral forms; with some products there has been a several-fold increase in yield, over 1% per century, and sometimes, as in wheat and maize, form has changed so much that natural ancestry is obscured. This increase in the predominance of the mean has been achieved at a cost of a decrease in variety, and doubtless with the loss of much genetic material. This is due to the strength of selection for a few obvious features encouraging the loss of alleles influencing others. This effect is amplified by the funnelling of small parts of the genome into later generations through a few individuals which dominate the ancestry of the future.

In the last 50 years cross-breeding, the development of novel strains by chromosomal selection, especially in wheat, and the 'male cloning' made possible through artificial insemination have further increased the rate of advance of the mean value of the essential product at a cost of further reducing the variance. This reduction in the variance, with its potential for future disaster, has recently become a desired feature, owing to the development of harvesting machines which can work efficiently only on products of standard shape and size, to the commercial advantages of packing items of regular size and simple shape, and to the associated development, and commercial encouragement, of a public preference for tidy displays of products of regular size, shape and colour. Our species has survived through parental education, and obedience to its dictates, assisted by ancestral memories of classes of edible plants, and is very susceptible to advertising that exploits aversions.

It appears to be assumed that this genetic advance can continue to increase production at a cost, now regarded as an advantage, in reducing the variance, while the environment can be made to fit through increasing reliance on artificial means of conferring immunity, killing pests, and artificially supplementing the normal constituents of food or soil. There are even international committees defining the legal limits of size, shape and content.

This separation of the genetic endowment from the environment in which we must live and in which our forests, fields and farms must operate is one of the more serious artificial divisions in both medicine and agriculture. It is consolidated in administrative structures, aided by unnecessary neologisms and, in genetics, often supported by an inappropriate algebra of such complexity that even the simplest operations are delegated to computers and protected from adequate scrutiny or quality control. Biometry, which was claimed to divorce and partition the joint contribution of nature and nurture, and was to have solved all the problems of agriculture, has largely declined as an intellectual force outside a very few centres of excellence which apply it with success to suitable problems, and some departments of psychology. The 'New Genetics' and its child, the Human Genome Project, are similarly at risk from numerical babelism although, as yet, the genetics of other species have been influenced little. These have greatly benefited from the development and documenting of various probes with cross-species variability, yeast technology, and other activities of a non-numerical nature.

There are no non-genetic disorders in man, beast or plant, but, as with infections and smoker's cancer, there is often a necessary factor external to the organism, allowing even the most susceptible members of the population the opportunity to lead a life free from this risk. The term multifactorial, much used in human genetics, has provided a superficial gloss to the misunderstandings of this interaction and increased the respectability now accorded to its study by means beyond the understanding of its more enthusiastic proponents. However, the term has little meaning because it applies to the 99% of illnesses which are not Mendelian or monogenic or unifactorial (a confusing triad of synonyms). Heritability (defined as that part of the phenotypic variability that is genetically based), which is usefully correlated with speed of response to selection in a constant environment and has an established place in applied agricultural genetics, has little relevance to the nature–nuture issue in a wild-type organism living in a variable environment such as ourselves. In particular, there is no justification for assuming any correlation between 'heritability', measured on a scale from zero to unity, and lack of opportunity for preventing disease, illiteracy or crime in most, but not all, individuals by environmental means. Height, intellect, athleticism and skin colour have high heritabilities, however measured or mis-measured, but the potential for growth of stature and amplification of intellect and physique is dependent on food, education and exercise, while the various hazards associated with skin colour, varying from sunburn and melanoma in the fair to rickets in the dark, are capable of being avoided or corrected.

Until the mutual dependence of nature and nurture, or nucleus and cytoplasm, is appreciated in both medicine and agriculture, and artificial divisions cease to be imposed by administration and consolidated by various misapplications of words and numbers, it will be difficult to bridge this widening gap, or to appreciate that short-term gains can be purchased only at an increase in long-term risks. These risks are potentially catastrophic and a century is a short time in genetics.

The last half century of scientific breeding has provided crops in a starving world beyond the wildest hopes of those who pioneered these advances, and bought time at a fully acceptable cost. However, this short-term investment, a sort of anti-insurance policy, is acceptable only if we admit that it is a major overdraft on future health and future investments for long-term health. Technologies are available for moving food across continents and piping water, including water desalinated by solar power, within continents. There is no feasible technology for changing the defence systems of major crops within less than a few years or those of farm animals in less than a few decades, especially as the chaos following catastrophic failures, which would probably be worst in the rich countries, would make both decision-making and funding even more difficult, particularly as those most competent to resolve the problem would be those who had created it.

The green revolution can wilt, and, worse still, will wilt, and will wilt suddenly. At worst, a billion deaths among the several billion who would not now be alive without these advances might in theory seem a small price to pay. In practice, such a disaster, apart from causing immediate suffering dwarfing any known past catastrophe in the 50 million years of mammalian evolution, is likely in the long term to lead to irreversible disruptions and an aversion to the continuation of established scientific procedures under the pressures of political biology and the increasing cloud of political correctness. The architects of this catastrophe, who have a monopoly on the techniques necessary for its resolution, would hardly be trusted to advise or execute the rescue of the bulk of our species from further starvation and its apocalyptic companions.

The production of high yields under optimal conditions is associated with a lack of genetic variation, due both to the products lacking variation and to their being few in number, so that vast areas of land are clothed, or grazed, by a few species whose members are unnaturally similar. These species are necessarily exposed to an increased risk of a single pathogen, and, in plants, an external predator, wiping out the whole variety. In the past the frequent milder and localized epidemics would leave survivors which necessarily seeded the future so that resistant stocks developed, without any knowledge of mechanism. There was stability as well as endemic disease. Now, such is the confidence in environmental control that it appears to be an established policy, when there are some survivors after an epidemic, to ensure the cycle of infections is broken by destroying the survivors, burning fields or shooting cows. This is simpler and cheaper than segregation and quarantine and avoids any acknowledgement of the long-term limitations of purely environmental control.

A further potential for catastrophe from some high-yield strains follows the use of cross-breeding between inbred varieties. In plants, the seed produced, if any, may be unable to breed or to breed true, so that economic catastrophes or disruptions of transport from strikes, lock-outs or civil war will deprive populations of the next season's crop.

There is nothing original in this, nor is there any lack of precedent. The Irish potato famine was the predictable consequence of reproducing a single crop from a limited ancestry. In South America and Mexico it was usual to grow potatoes in the company of wild and inedible varieties, so conserving variability at a cost in yield, a habit once considered symptomatic of agricultural ignorance. Without the potato most of those who died in Ireland would not have been born: however, death after birth is not, in a secular world, commensurate with non-birth.

There seems no easy solution. There is no justification for regarding the magnitude of the success of the green revolution as grounds for assuming that the expected steady increase in yield can be achieved without the risks of catastrophic global losses. One possible solution would be a requirement that high yield seed should be mixed with 5% or so of variable and fertile seed. This would reduce the yield by a small proportion, but, in the event of an epidemic, it would be unlikely that there would be no survivors, and, in the event of a failure to deliver seed for the next season, there would be something to grow.

The genetic problems of low variance and clonal foods resemble those of nuclear power—the benefits are undoubted, provided no catastrophic event occurs and the technology of the future can deal with the various long-term consequences. However, apart from a recent interest in slash-and-burn methods and wide-scale logging, catastrophic agricultural events are considered unlikely.

Expert opinions will differ in the time-scale and magnitude of the risk. However, there is no doubt that there is a risk, and such risks need the insurance of funding both in research and in the storage of wild and old varieties in the public domain, with more than one repository for seed, and with adequate resources to grow and harvest at appropriate intervals.

Direct evolutionary consequences for humans

The genetic problems of our own species are somewhat different from those of domestic animals and crops. Nothing sudden can be foreseen, but slow and irreversible damage can easily be imposed on a world-wide scale through the dissemination of pollutants by air or water or by the massive and widespread commercial distribution of inadequately or inappropriately tested pesticides, food additives and novel foods. Even if nothing is done it is not clear what scale of differential reproduction related to those features which distinguish us from the higher apes is necessary to maintain the *status quo* in our ability to inhabit and sustain a civilization worthy of the name.

It is simplest to visualize our species as one which has distinct generations, like annual plants and most butterflies: this clarifies without disturbing the underlying balance sheet. We can consider the heritage of one generation basically as a box filled with genes, either as a mere sac or 'bean-bag', or with the ultimate genetic elements, the three thousand million base pairs which make up our gametes, lined up on their chromosomes, with the chromosomes laid end to end horizontally, each individual (because each person is compounded of two gametes) having a pair, with the individuals being stacked vertically.

There are a lot of elements, and even more individuals. It is the responsibility of the present generation to ensure that it exports to the future a set of elements which is not inferior either in numbers of seriously detrimental elements or from loss, or corruption, of those subtler elements which distinguish us from our ancestors of a mere 200 000 years or so, who we may reasonably assume lacked some of the abilities by which our lives can be enriched and which provide the necessary basis for a civilization.

Before these statements are interpreted on inappropriate grounds as 'politically incorrect' (which they probably are), perhaps I could consider the analogy of a living language—Dutch, for example. I assume that those who speak it value it, and accept that it has improved in the last 200 000 days, or 600 years. Most native speakers will wish to protect it from excessive mutant forms with a view to its retaining its power as a civilized structure for at least another 600 years. Considerable resources amounting to several hundred hours of formal and several thousand of informal instruction per child is regarded as a sound investment by parents and tax-payers, and the concept of good and bad speech and writing is not regarded as divisive or disruptive of a sound society.

In this simple analogy we might consider the living language, as preserved in speech. If 14 000 000 people speak for two hours per day at 200 words per minute there will be several thousand million words of some hundred thousand different varieties, very few of which will be unique and most of which will be repetitive elements. Any day-to-day inventory of words will define the elements, though not the structure, of a language, and, unless words are passed from today to tomorrow by usage, they will die from neglect to be replaced by superfluous neologisms, incongruous mutant forms and unwelcome immigrants. This would be followed by nostalgia for a dead language accessible to only a declining minority. Unacceptable words, once incorporated, will have a substantial half-life and can die only from neglect. The natural tendency to decay is halted by education and the creation of a desire to speak well, and, for an influential minority of any population, to write well. This is well understood but not easily amenable to further clarification.

The transfer of the human genome from generation to generation is not entirely lacking in analogy. It is unlikely that anyone would dissent from the view that it is too valuable an asset to allow it to be corrupted and to decline, although the meaning of decline is difficult to define and attempts made in the past to both define and halt 'decline' have been barbaric, catastrophic, and short-lived. What is clear is that as a society we need to maintain both our mean abilities in such fields as our society needs, and our variability in these abilities, because the natural variation in innate aptitude is such that the diverse needs of society in both industry and the arts cannot be satisfied by educating individuals selected without reference to this. This is perhaps more obvious in the field of athletics, in which there must be some selection by shape, size and ability to exert either sudden or prolonged effort for the various events.

In considering the implications of this aim to transfer from our generation to the next as few harmful elements as possible, so that matters do not become worse, without losing the largely latent variability on which the future of our species depends, we have to take into account those matters within our understanding and also accept that there are others.

In language, the simplest measure is the 'word count' and some words, or, by analogy, genic variants, we can agree we would be better without. In genetics, this includes most determinants of severe Mendelian disease but probably excludes most predispositions to most other disorders of body or mind. If the future is good, we might be able to dispense with the dangerous insurance against malaria provided by sickle-cell disease and related disorders which are the major causes of death from single-gene variants; we could then in practice dispense with the alleles which protect the carrier. The premium presently paid for this protection is several per cent of deaths in childhood in some countries. Fortunately for the long term, birth avoidance either by discouragement of some marriages or by fetal diagnosis and abortion and death before birth will actually increase the proportion of carriers who are relatively protected from a disorder which is likely to resist complete extermination for some centuries of consistent peace in areas where it is endemic: if global warming occurs, these areas will increase.

Disorders common enough to be catalogued and simple enough to be detected before marriage or birth are only a minor part of the problem. At the single gene level of Mendelian disorders the new disorders caused by new mutations will not be manifest unless they become established through chance before classification and the technology of diagnosis is established, by which time there will be thousands if not millions of copies. The single gene problem is itself minor. Most of the problems of both health and harmony in a complex society are compounded by the various combinations of numerous genetic elements which predispose to ill health, poor fetal development or poor intellectual developments even within the best domestic, nutritional and educational environments, and propensities to violence, fraud and other variant behaviours which will always require the coercion or isolation of a minority, and all that this entails. That there is an inborn variation to criminal activities should hardly be in doubt; although such activities are greatly encouraged by the complex of poverty, ill health and boredom, it is at least clear that almost all criminals, especially the more violent variety, are male, and maleness is an undoubted genetic difference. There are problems, and these are not eased by denying that they exist.

The ways in which society supports health and education, and influences the opporunities of its members in housing and other aspects of the provision of a reasonable environment for children are not without impact. A very simple example is the mortgage rate, which, if high, inhibits the reproduction of that subgroup of the population who need mortgages, in which those with professional qualifications are over-represented. These matters are obvious and only obfuscated by theoretical analyses. The uterus is at its prime at twenty but the environment outside is usually not. If we consider the simple problem of cleansing the genome of clearly undesirable elements from past mutations, the enormity of the task becomes apparent. Tay-Sachs disease, for example, is a recessive condition with an incidence of about 1:600 in Ashkenazi Jews and perhaps 1:50 000 in most communities. This implies that the proportion of genes with this mutant is about 1:40 and 1:800 respectively, in a world with seven billion individuals and seven million widely dispersed Ashkenazim. Only about 2% would be in a population in which screening would be economic or, because an error rate of at least 1:800 is to be expected in any biological screening, free from excessive anxiety which leads to further costs in double-checking and erroneous termination. For many centuries there will clearly be better ways of investing in health on a world level, as opposed to a local level, in communities in which sufficient prosperity, medical standards and risk of affliction are combined. Institutional care is no longer politically correct—abortion is.

We cannot cleanse the genome. Eviction before birth resolves immediate problems, and has been done with humanity and efficiency in societies in which resources and opportunity coexist, for Tay-Sachs disease, haemoglobin disorders and, recently, with cystic fibrosis. However, the long-term result is to very slightly, but progressively, increase the proportion of disastrous mutants. Any attempt to cleanse the genome by discouraging the reproduction, or birth, of all carriers would leave only a small minority of potential parents, or of surviving fetuses. The loss of 90% or so of pregnancies on grounds of the long-term health of society is hardly practical. Perhaps this is fortunate, because judging the unborn is more difficult and open to abuse than any other judicial enterprise: the defendant is absent and there is little time for retrial, and expert witnesses are confused and their terminology confusing. Delegation of risk analysis to computers has been tried, and is still used, but it has come into well-deserved disrepute in the forensic world after wrongly assigning criminal guilt in adults by inappropriate procedures.

We cannot cleanse the genome. What gets into it through damage by mutation can be removed only at a cost of fetal death or, what in most societies is regarded as the far greater loss, death after birth, and this will be after a delay of one generation in dominant disorders and over many hundreds in other disorders. Diagnosis of carrier status can only relate to some rare and usually geographically and ethnically restricted disorders, and any action will usually increase the representation of most mutations in the next generation, even if it limits the distress they cause in this generation.

All that can be asserted with confidence is that there are catastrophic genetical opportunities to disrupt our society, and it is for experts to assess the magnitude or imminence of these risks. So far as medical intervention in genetic disease is concerned, what is being done is, on the whole, done wisely and humanely within the restraints of resources and the education of those competent to use them. The long-term effect of intervention is mainly detrimental, slight but progressive. There is no cause for alarm in that: before anything can be done for good or ill we shall be better informed to evaluate it.

Meanwhile, all that we can do is ensure that exposure to mutations is as low as possible, and that undue attention is not directed to the best understood of the environmental mutagens, radiation. It is at least clear that this is a minor contributor to the normal mutation rate, outside accidents on a scale not yet evident or catastrophic decisions about the production and handling of waste. Even in the Ukraine most mutations over the last decade will have been induced in other ways. Alarmist reports can only divert attention and funds from the more serious hazards of what we eat, drink and, if smokers, breathe. Genetic counselling will have little influence on future gene frequencies, and what little effect it does have will often make the future worse. However, it is urgently needed in the vegetable kingdom.

DISCUSSION

Kleinjans: Toxicologists are trained to see mutations, chemical or physical damage of DNA, as harmful, and I suppose that will be correct, considering, for example, oncogene mutations and so on. With chemical or physical DNA damage, however, mutations will occur all along the genome, and some of them might be beneficial. Could you speculate on the ratio of beneficial to harmful mutations?

Edwards: The odds of a beneficial effect must be very remote, perhaps 1000:1 or 100:1, in any fairly well-tuned system. There's a lot of variability anyway, although there have been grossly exaggerated claims about the number of defective genes which would have serious effects if they met each other, causing recessive disease leading to death after birth. On average, each of us probably carries at least one but probably less than two disastrous genes and a lot of undesirable ones. I think we have enough variation to be getting on with, and would be well advised not to introduce unknown novelties, although we obviously could not have evolved without variation.

Mansfield: You said that we depend on a small number of animal and plant species for food. As we experience environmental change, occurring more rapidly perhaps than ever before, are we going to be limited in our ability to modify those species genetically so that they can inhabit different environments? Some of the requirements will be met by taking genotypes that are adapted to perform well in one part of the world and moving them to another part of the world, but it won't always be simple to do that; the frequency of extreme events is important, as is photoperiodism.

Sutherst: There are certainly a limited number of genotypes in domestic animals. My advice to people in Africa who are sent Friesian cattle is to eat them when they come off the aeroplane, while they are still alive, because they're poorly adapted to such an environment. My feeling is that climate change itself will have a limited direct effect on the productivity of the animals, but that the effects will be exerted through effects on vegetation, the protein content of plants, and through pests.

Mansfield: I was thinking particularly about plants, where there are more limitations than many people realize. It troubles me to see people showing maps of the expanding areas in Europe in which it will be possible to grow grain maize, because we don't know anything about the future occurrence of extreme events. If there is a rise in mean temperature of 2 °C, but there is still a danger of frost in late May over much of north-western Europe, I don't think there will be any change in the areas in which we can grow grain maize, unless we succeed in manipulating its tolerance of cold.

Lake: It won't survive and grow with temperatures around its roots lower than about 5 °C.

Edwards: The changes in weather will not happen very quickly, but changes in numbers and varieties of viruses and aphids will. The entire Brazilian economy is based on clonal coffee, which could be wiped out by a beetle in a year or two. The situation is the same with grape vines; even the variants are somatic mutations. This is rather a minor problem in comparison with wheat. In the short term, it is important that there should be facilities for growing seed material within the country in which the wheat is grown so that the country is not totally dependent on imported seed.

Sutherst: When we try to infer geographical distributions of crops or wild animals and plants, we have to be careful about how we derive those estimates. The more I look at geographical distributions, the more I admire the way in which organisms integrate all the different factors. A geographical distribution is not determined by the average conditions but by the average together with the variability at each location. Variable conditions result in selection for different characters in different years. We've had problems in Australia when geneticists have tried to select for a particular character during a succession of wet years followed by a number of dry seasons during which other characters were advantageous. Extreme events will certainly be important. What we don't yet know is how much the *variability* of the climate will change. The extreme events probably limit the geographical distributions of plants more than the average conditions do.

Edwards: A cow will survive one frost, but many plants will not.

Mansfield: In the case of grain maize, one frost is sufficient to damage the whole crop.

Sutherst: In some areas in Australia which are presently suitable for growing stone fruit (e.g., plums and peaches), these crops are being abandoned because the high frequency of hail makes it uneconomical to continue. Unlike cultivated crops, wild plants can't be replaced after an extreme event has killed off the local population, hence my concern for natural ecosystems.

Klein: Genomic diversity was an issue at the World Summit held in Rio de Janeiro in 1992. How much are we as humans really able to manipulate this parameter? A relatively small number of species are involved in agriculture. How will global change affect world-wide genetic diversity in a way that we cannot control? Do we have any measure of diversity which is not taxonomic, not species dependent? Can the geneticist suggest a rational approach to genomic diversity, without looking into one species?

Edwards: The best indication of diversity is that a supermarket manager won't buy a product, and the simplest measures of diversity are things like size and shape. In suppressing variation in size, shape and colour, other variations are suppressed. However clever biochemists are, they are not as numerous as bacteria and their experiments don't have such a good end-point. With bacteria and aphids and others developing new enzyme systems, chemical companies are no match. They can win in the short term, but in the long term it's going to cost more and more to maintain the environment in which the clonal organisms can grow. That is really inevitable.

The only hope when there is an epidemic is that something survives. The survivor should not be killed, as under present policy with respect to foot and mouth disease, for example, but nurtured and allowed to reproduce. That's what happens in Nature. Take wheat and wheat rust as an example. They are continuously at each other and have been doing this for so long that they've virtually agreed that 10% of the wheat will be lost in order to maintain coexistence. If you take away the rust completely, another set of problems will develop, and if the rust comes back it will do so with a vengeance.

Zehnder: The most simplistic way to describe biodiversity is the sequence of genes of all organisms. However, simple sequencing is not sufficient to describe biodiversity. Modern taxonomy does not just describe what an organism looks like, but combines the organism's physiology with its biochemistry, ecology and genetics.

Bobrow: Professor Edwards, you painted a picture of global catastrophe in which the world's wheat is wiped out because of its lack of variance, and said that even though the possibility of this happening is remote, its consequences are so severe that some form of insurance seems reasonable. That's very persuasive, but I don't understand what sort of insurance one could take out against it. You gave the example of a cow which is killed after surviving a viral infection, but that cow has survived a *known* pathogen, whereas the real problem you are indicating is a presently *unknown* pathogen against which we will not be able to defend ourselves. What sort of strategy could one adopt in order to provide the sort of insurance that you are talking about, other than deliberately avoiding the breeding of even partially standardized crops.

Edwards: One approach would be a requirement for imported sterile or simple cross-bred seeds to include 5% of variable fertile seeds. These would be a nuisance, of course, because they would be too big for the combine harvester, or too short, or would get in the way, or get eaten or would introduce rusts; this would be like an insurance policy—it's wasted if nothing happens—but if the worst came, and 90% to 95% of the main crop were wiped out, the survivors would be resistant. The survivors would be a stock from which one could breed. This would also help with local problems. Consider the situation which has arisen in Yugoslavia. It is impossible for seed corn to be imported; I don't know what corn they have, but it may well be impossible for them to grow any corn because it is sterile. Such a situation can happen anywhere, not only as a result of wars but also because of hurricanes, or disruptions of ports or railway lines. I don't know what the ratio of corn to seed corn is. I suppose one wheat grain produces a plant with 50 or so. Rather than starving if your supply is disrupted, it would be better to have some seed corn left. All sorts of catastrophes have a 1% chance of happening in the next decade. I would have thought that even economic disintegration in the USA had a risk of one per thousand per year. I doubt if even the most optimistic banker would insure for less than that. The complete disruption of the USA as a viable financial structure which is able to run an export service would be a major catastrophe, because it's the most advanced and efficient exporter of corn. As with nuclear power stations, you have to think ahead and prepare for the worst, however unlikely.

Sutherst: I don't think we should get the idea that these sorts of catastrophes are unlikely. We have recently witnessed the almost total loss of a corn crop due to corn blight in the USA. At the moment there is an uncontrollable outbreak of insecticide-resistant white-fly in the USA which is wiping out many different crops.

Final general discussion

Zehnder: Martin Bobrow suggested (p 233) that human health is the most important thing. This suggestion is not entirely in accordance with our discussions after John Edwards' paper. Is the anthropocentric view, that human health is all we are concerned about, the right approach, or do we think that Nature around us also has a certain right in itself to proliferate?

Bobrow: I too feel concern for other biological systems, but they are secondary. My worry is not primarily that wheat might die out as an organism, but that if the wheat crop is wiped out there will be a lot of hungry people around. Where I think there is a genuine disagreement is on the issue of whether the end point of control is to achieve some sort of clean environment with reference to a historical standard, or whether it is to minimize measurable (or at least reasonably conjectured) biological effects. I'm concerned about the extinction of an animal species, but more so about the wiping out of humanity. I am not comfortable with the idea that huge amounts of money should be spent removing the last residual traces of some pesticide from the water supply or soil in a rich country, simply because it is not 'natural' if this will reduce the resources available to combat the appalling human problems which David Bradley and Dennis Lincoln described.

Klein: One could, for the sake of argument, ask whether there is any need to worry about human health. Human population is increasing incredibly, life expectancy is increasing in most countries, and no one dies healthy—life has to end in one way or another. Statistically, human society has never been more healthy than it is today. There are other species on earth which may, in the long run, be more important for what we call a living and sustainable ecosystem. We have to be careful to avoid ranking the human position too highly in this discussion. Considering genetic diversity and the living world as a whole as far as we can, we have to ask ourselves what effects man-made global change will really have on the biological world as a whole.

Avnimelech: As we are approaching the end of this symposium we should try to see the whole picture, but the whole picture is complicated. On the one hand, we have been told of the increasing population and the increasing demand for food and for shelter. There is no doubt that even the most sophisticated technology and the most intensive development will not be enough to feed the future population. Biological diversity has to be combined with the development of the most productive crops if we are to feed the population. We have limited resources; there are 800 million people with enough resources but many billions without. At some far point in the future, if human beings survive after the period of population explosion, we may be able to worry about not increasing cadmium in the soil and increasing population diversity and genetic diversity. In the next few decades or centuries, we will have to optimize our responses—this is what we as scientists, as people who make or influence decisions, have to think about. What is the best strategy that we should adopt? Maintaining human existence is one goal. As David Bradley pointed out (p 244), to solve the problems of the Third World, there is a need to return to basics, rather than extrapolating or interpolating from the developed industrial world. To get the whole picture, we have to use parts of our scientific knowledge and experience assembled differently under different conditions of finance, attitude and culture. We cannot look at isolated aspects alone. We have to take blocks from all of the presentations made here to try to come up with an optimal solution.

Kleinjans: It's too simplistic to say that the human population is doing well because it is increasing. We experience human health at an individual level. We cannot avoid taking an anthropocentric position. We cannot read the rights of Nature in a book produced by Nature itself—we have to use human words and human concepts to describe whatever we think Nature's rights may be. There's nothing wrong in taking a human health parameter as a calibration point for environmental policy; you should not isolate it as a goal, but you can use it as a calibration point. What you have to do is to spread the benefit of that policy equally over the human population, Third World countries included.

Mansfield: I would like to raise a slightly different issue, but it does link up with these themes. We have discussed the effects of environmental change on agricultural production, but we have not actually discussed the overall importance of the process of photosynthesis to human existence. Listening to what Dennis Lincoln said made me feel that that whole process is really endangered, because if our objective is to 'improve' the lifestyle of five or six billion people, and if that so-called improvement brings about roughly proportional deteriorations in the aerial environment, for example, an increase in tropospheric ozone concentration or further stratospheric deterioration so that more UV penetrates, the process of photosynthesis will be damaged on a global scale. This will affect CO_2 exchange. If CO_2 intake by plants is reduced, the rate of increase of CO_2 in the environment will increase, and global warming will increase. Also, when plants begin to be damaged in this way they tend to increase their respiration, because repair mechanisms increase, so this will enhance the whole effect. We might get a steepening of global warming, and one can see the situation getting out of control. This brings us back to the point about our first priority being to secure human existence. That would be a situation in which our main concern would not be about the health of individuals, but we would be concerned about the continued existence of all forms of life on earth.

Sutherst: I am concerned that we keep talking about humans as if we're something special and that the environment is a resource exclusively for our use. We talk about water in Europe, and if something in the water doesn't affect humans we say it doesn't really matter. I would be surprised if the concentrations of nitrates and phosphates in polluted European water wouldn't kill a large proportion of Australian native plants, for example. A lot of animals and plants are sensitive to things that we are not sensitive to.

The feeling I had yesterday when we were discussing nitrates in drinking water (p 215) was that concerns about water were dictated by the cost of cleaning it. Whoever causes the pollution should pick up that cost. If the polluters want to pay to clean the water so they can keep polluting it, that's something for them and their peers in the cities to consider.

My main concern is that we talk about human health while we should be talking about global health. What's good for owls is also good for humans. If you look after the general environment in it's widest sense, human health will in many ways be taken care of. A more sensible approach to human problems might be to treat our own species scientifically, just as we would any other species in the world. Our consciousness and awareness of self interest prevent that and will probably be the cause of our failure to control human populations.

Klein: The philosophy of taking chemical elements or compounds back to where they came from is contradictory to human behaviour. It is a key feature in human culture, but not that of other animals, to purify and recombine elements to make new materials with new functions. We are mining and concentrating materials in a way which is not natural, and it's impossible to redistribute the elements in a natural way. It may be feasible in a few cases for whoever brought about the separation or enrichment to redistribute or dilute the elements back to the natural origin, but this concept is not realistic and is not helpful as a general policy.

Edwards: Human health must include mental health. An environment without any trees or music wouldn't actually affect the death rate but would affect what one could reasonably call mental health. The problem is resolved if it is accepted as environmentally desirable that we shouldn't exterminate the elephant just for convenience.

Bobrow: Several people have made similar points from very different points of view. John Edwards just stated that when one talks of humans and human convenience, one is talking not only of quantity but also quality of life, and I would entirely agree with this. Furthermore, man is part of a complex ecosystem, and therefore the well-being of humankind is largely interrelated with the well-being of many other species. However, in the end, choices have to be made. If you take the point that Bob Sutherst made to its logical extreme, you would have to say that you ought to be as concerned about the survival of Australian lake plants as you are with human survival. You then have to translate this into practical guidelines for legislation. If your baseline is that all species are of equal importance, are you going to apply that to parasites? Are you going to say that *Onchocerca* has an absolute right to existence, and that all attempts to poison human pathogens are a diminution of biodiversity? Although this idea sounds attractive, we have to accept that some of these things are more important than others. Having a rank order of importance doesn't mean that everything else is unimportant. It's not that I don't care about Australian lake plants, but I just care rather more about some other things, and if I have to make choices, those choices will be dictated by my priorities.

James: We as environmentalists, if I am allowed to call ourselves such, are appealing to society for funds and for action in a mode which is often appealing to common human emotions and concerns about survival. It's true that there is another view, a philosophical view, as we have just heard, but we have been discussing a whole array of different issues and it is rather difficult to generate the sense of a crisis. We've heard about what might happen if there is a 1% chance of eliminating a major crop. That seemed to be a real problem, because we currently can't feed ourselves effectively. Then Bob Sutherst tells us that we are actually being completely stupid, that people are pests who grossly contaminate and change the globe. As a physician, I am worried about the suffering of society. I think that we end up arguing only about priorities in terms of spending and organization of society. If we accept that we are dealing with priorities, then decisions have to be made as to whether we should put billions of ECU to bring salmon back to the Rhine, or to eliminate copper contamination of soil in The Netherlands, or to provide technical aid to assess the problems in the Third World. In the mean time, as Dennis Lincoln has stressed, we should get cracking on population growth, otherwise all these sophisticated analyses are going to be overwhelmed by the one big problem of overpopulation.

Summing-up

Alexander J. B. Zehnder

Swiss Federal Institute of Technology (ETH) Zürich, Federal Institute for Water Resources and Water Pollution Control (EAWAG), Ueberlandstrasse 133, CH-8600 Dübendorf, Switzerland

I would like to highlight some of the issues we have discussed over the last three days and put them into a broader perspective. The enhanced greenhouse effect is real, and, according to the knowledge we have today, will trigger global warming. However, our predictions are based on physical models. Scenarios incorporating chemical and biological responses seem to be highly complex and remain hypothetical. These responses have therefore been only marginally included in models of global warming. Considerable research needs here are obvious.

Globally, pollution of air, water and soil steadily increases. Negative effects of pollution on human health have been identified. Encouraging local results have been achieved in combating pollution, in particular in lakes and river systems of some Western countries. These examples show that special efforts can produce positive results. Cleaning up and restoration of polluted compartments requires decades. A steady pollution has taken place over years, and years will be needed to cleanse the sins of the past. Actions taken now, however severe, may only show results in five, ten, twenty or more years. Success is not guaranteed for all actions. For most of mankind's activities, zero pollution is not possible. Therefore, quality standards and controls have to be applied, but these are strongly debated issues. Protection and sustainability of our resources can be achieved only if in addition to scientific arguments short-term and long-term socio-economic aspects are taken into consideration.

Global warming and pollution may become threatening for mankind in certain areas over the course of the next century; however, this is overshadowed by the tremendous population explosion. We can expect world population to double in the coming 50 to 80 years, reaching the 10 billion mark by 2050. Extrapolation models predict that an optimized agricultural system could feed world population in the middle of the next century, but the main areas of agricultural production are different from the main areas of population growth. We shall be faced with a distribution problem considerably more severe than we are used to now. Global warming and intensified agricultural production will allow major pests and diseases to move into places where they have never been before, triggering an increased use of pesticides which in return will threaten essential resources such as ground-water. Most programmes for nutrition are based on models with considerable shortcomings. One may ask, what are the consequences if our future models are not correct? The scientific community is surprisingly silent on some of these issues.

A special effort is needed to prevent a further degradation of the environment, ideally without the loss of the comforts we are used to. It seems impossible to combine further industrialization, increased agricultural production, the desire for better individual lifestyles and mobility, and improved personal safety and comfort for a fast growing number of people, while still protecting our environment. However, if we are to hand down to our children an intact world and a quality of life similar to ours, such a seemingly impossible combination is exactly what is necessary. The key is sustainable development. This is easier said than done, because local, regional and world-wide concepts are still missing. Sustainable development is a challenge for us all, regardless of our occupation and position in society.

During this meeting some doubts were expressed about whether politicians and policy-makers have the competence to make the right decisions; however, we did not really question our competence to supply them with the information necessary for them to make responsible decisions. At this meeting, we have found that our answers to questions were often not straightforward; we had to make qualifications, to say that existing information does not allow clear-cut statements, or that the problems were too complex to be covered by a simple answer. How can we expect policy-makers and politicians to make competent decisions on the basis of vague and debated information? We scientists have to learn to separate much better the crucial, generally applicable conclusions from the less important ones. This would greatly improve our communication with the media. The outside world is interested in clear and relevant statements and not in our complications, hesitations and 'on the one hand but on the other hand' arguments.

I hope this symposium has given and will continue to give you more insight into the questions addressed by neighbouring and remote disciplines and will allow you to see your own work and your own discipline in a broader perspective. I sincerely hope that you will make use of the knowledge and skills of the other disciplines for your own research and formulation of your research objectives. The rapidly increasing number of central questions which have to be answered in the coming years need our full attention. Only a close collaboration between disciplines will enable the vast problems with which our future world is confronted to be tackled successfully. We already have to deal with the problem of ensuring that our children and their descendants have a quality of life comparable with ours. Quality of life includes not only nutrition and shelter, but also an intact ecology, intact resources and an intact biodiversity. We must preserve and hand down to future generations Nature's capital. I am sure that this symposium will contribute to our efforts to reach the common goal—sustainability.

Now it remains only for me to thank everyone who has helped to make this symposium a success.

Index of contributors

Subject index

Abortion, induced, 202
Acid rain, 30, 40
Acidification
  atmospheric, 27, 30, 39–40
  soil, 122
Acquired immune deficiency syndrome
  (AIDS), 101, 137, 204–205
Action Programme Rhine, 56, 59, 60, 61
Aflatoxin, 34–35
Agent Orange, 48
Agriculture, 264
  catastrophic events, 251–252, 258–259
  developing countries, 79, 203, 207
  greenhouse effect and, 62–79
  Western/European, *See Western/
    European agriculture*
AIDS (acquired immune deficiency
  syndrome), 101, 137, 204–205
Air pollution, 31–32, 36–37, 39–40,
  209–210
  cleaning up, 35
  crop yields and, 40, 99–100
  health effects, 33–34, 171–172, 173,
    179–180
  insect susceptibility, 170
  past, 24–25
  present, 26–27, 28
  private vs public, 37
  See also Smog
'Airport malaria', 161–162
Alcohol, 169, 231
Ammonia (NH$_3$), 18, 30
Anaemia, 83–84, 164
Animal protein
  consumption, 84, 97
  production, 88–89, 98, 102–103
Anti-gestagens, 202, 212
Anti-schistosomal drugs, 163
Arctic haze, 27
Arsenic (As), soil content, 110, 111
Arthropods, as disease vectors, 124–145
Asbestos exposure, 226–227

Asbestosis, 226–227
Asthma, 179
Atmosphere
  acidification, 27, 30, 39–40
  oxidizing potential, 38
  See also Air pollution; Greenhouse gases

Bed nets, insecticide impregnation, 142
Benzo(a)pyrene (BaP)
  lung cancer risk, 173
  molecular markers of exposure, 174,
    176
Bioavailability, compounds in soil,
  115–117, 118–119
Biogas production, 243–244
Biological control, pests, 99, 101, 144
Biological oxygen demand (BOD), 26
Biomarkers, 172–173, 177, 181
Biomphalaria snails, 238–239
Black-flies (*Simulium*), 124, 125, 126, 238,
  239–240
Blood pressure
  INTERSALT study, 222–224, 225,
    229–233
  urbanization and, 223, 232, 244–245
Body mass index (BMI), mortality and,
  92–94
BPDE (7,8-dihydroxy-9,10-epoxy-7,8,9,10-
  tetrahydrobenzo(a)pyrene), 174, 176
Breeding, selective, 248–252
Bulinus snails, 238–239

C$_3$ plants, 66, 67
C$_4$ plants, 66, 67, 69
Cadmium (Cd), 31, 169, 170
  acceptable contents in soil, 113–115,
    116
  carcinogenicity, 178
  reference values in soil, 110, 111, 112
  River Rhine, 45–48, 60
Cancer
  electromagnetic radiation and, 183

267

28 air quality
39